Lecture Notes in Control and Information Sciences

Edited by M. Thoma and A. Wyner

For information about Vols. 1-96 please contact your bookseller or Springer-Verlag

Lecture Notes in Control and Information Sciences

Edited by M. Thoma and A. Wyner

173

W. Krabs

On Moment Theory and Controllability of One-Dimensional Vibrating Systems and Heating Processes

Springer-Verlag Berlin Heidelberg GmbH

Series Editors
M. Thoma · A. Wyner

Advisory Board
L. D. Davisson · A. G. J. MacFarlane · H. Kwakernaak
J. L. Massey · Ya Z. Tsypkin · A. J. Viterbi

Author
Professor Werner Krabs
Fachbereich Mathematik
der Technischen Hochschule Darmstadt
Schloßgartenstraße 7
W-6100 Darmstadt
Germany

ISBN 978-3-540-55102-7   ISBN 978-3-540-46696-3 (eBook)
DOI 10.1007/978-3-540-46696-3

© Springer-Verlag Berlin Heidelberg 1992
Originally published by Springer-Verlag Berlin Heidelberg New York in 1992

The use of general descriptive names, registered names, trademarks, etc. in this publication does
not imply, even in the absence of a specific statement, that such names are exempt from the
relevant protective laws and regulations and therefore free for general use.

Typesetting: Camera ready by author

61/3020-5 4 3 2 1 0   Printed on acid-free paper.

Preface

One of the historical origins of moment theory which goes back to
the end of the nineteenth century is the problem of reconstructing
a probability distribution from the sequence of its moments. This
is a typical infinite moment problem in the space of functions of
bounded variations with infinitely many equations to be solved.
Later this problem was given an abstract form as a finite moment
problem in the space of bounded linear functionals on a normed linear space and was investigated in this form by M.G. Krein.

In the late fifties and in the sixties of this century it was dis-
covered that finite moment theory can be applied to finite dimensional linear control problems in order to obtain all the fundamental results concerning time-minimal linear control problems. The
first four sections of the introduction of these lecture notes are
devoted to demonstrate the application of finite moment theory to
finite dimensional linear control theory.

The application of infinite moment theory to infinite dimensional
linear control problems began in the late sixties of this century
and was continued in the course of the seventies with pioneering
work by D.L. Russell and H.O. Fattorini. Due to the fundamental
assumption that every finitely many moment equations chosen out of
the infinite sequence of such should be linearly independent this
application has to be confined to distributed or boundary control
problems with space dimension one. There are a few exceptional
cases of higher space dimension where infinite moment theory can be
applied but a convincing general approach is only possible for distributed or boundary control problems of space dimension one.
Therefore these lecture notes are confined to problems of this type
where in the case of vibrations a moment theory in Hilbert spaces
is applied that was developed late in the seventies by J.F.
Korobeinik.

The same results can also be obtained by the theory of linear operator equations which in addition renders to be very fruitful for time-
minimal control problems in connection with minimum norm control
problems.

For boundary control problems in heat diffusion exponential moment
problems have to be considered as a special case of a Banach space

moment theory. Here the theory of linear operator equations can also be used in order to reduce time-minimal control problems to problems of minimum norm control.

This text is based to a great deal on the authors own investigations and was also partly used as material for advanced courses in control theory.

Gratitude is owed to Mrs. A. Garhammer for her careful typewriting of the manuscript.

Darmstadt in September 1991

W. Krabs

Contents

0. Introduction: Finite-Dimensional Linear Control Systems and an Outline of Infinite-Dimensional Linear Control Problems.

0.1. The Problems of Null-Controllability and Null-Reachability.

The main goal of these lecture notes consists of the investigation
of linear control systems which are governed by partial differential equations and therefore are infinite dimensional systems. We
will be concerned with two kinds of problems: The problem of null-
controllability where we ask for a control by which the system is
steered from a given initial state to the "zero-state" (which, for
instance, is the state of rest in the case of vibrations) within
some given time interval and the problem of null-reachability where
we start with the "zero-state" and look for a control which steers
it to some given state within a given time interval.

The essential tool for dealing with these two problems will be
moment theory for infinitely many moment equations. Special problems
of this kind have been investigated in the late nineteenth and early
twenteeth century already (see, for instance, [1]). The application
of infinite moment theory to linear control theory, however, has not
been started earlier than the last three decades. It was preceded
by the application of finite moment theory to finite dimensional linear control systems. Possibly the first pioneer in this field was
Krasovskii (see [8]). Important contributions in this direction were
also made by Antosiewicz in [2] and Marzollo [9]. In [6] moment
theory is consequently applied to finite dimensional linear control
systems. Minimum norm moment problems are considered and are used
for solving the problem of null-controllability by norm-bounded
controls in minimal time.

The first aim of this introduction therefore is to demonstrate the
application of finite moment theory to finite dimensional linear
control systems. For this purpose it suffices to consider autonomous
systems which are governed by linear vector-differential equations
of the form

$$\dot{y}(t) = Ay(t) - Bu(t), \quad t \in \mathbb{R}, \tag{0.1}$$

where A and B is a (constant) real $n \times n$- and $n \times r$-matrix, respectively, and $y = y(t)$ and $u = u(t)$ is a real n- and r-vector function, respectively. The function $u = u(t)$ is the control function
and is allowed to vary in the space $L^p([0,T], \mathbb{R}^r)$ of all r-vector
functions whose components are in $L^p([0,T])$ for $p \in [2,\infty]$ where $[0,T]$

with $T > 0$ is some given time interval. Let $y_0 \in \mathbb{R}^n$ be fixed as initial state of the system. Then it is well known (see, for instance, [5]) that, for every $u \in L^p([0,T], \mathbb{R}^r)$, there is exactly one n-vector function $y = y(t)$ whose components are absolutely continuous and which satisfies (0.1) and the initial condition

$$y(0) = y_0. \tag{0.2}$$

Moreover, it can be explicitly represented by the formula of variation of constants

$$y(t) = e^{tA} [y_0 - \int_0^t e^{-sA} Bu(s)ds] \tag{0.3}$$

where

$$e^{tA} = \sum_{k=0}^{\infty} \frac{t^k}{k!} A^k, \quad t \in \mathbb{R}, \tag{0.4}$$

is the matrix-exponential function (with $e^0 = I = n \times n$-unit matrix). After these preparations we come to the

Problem of Null-Controllability: Given $T > 0$ and $y_0 \in \mathbb{R}^n$, find $u \in L^p([0,T], \mathbb{R}^r)$ such that the unique solution $y = y(t)$ (given by (0.3)) satisfies the end condition

$$y(T) = \theta_n \quad (= n \times n\text{-zero vector}). \tag{0.5}$$

On using (0.3) and the fact that the matrix exponential function (0.4) is non-singular for every $t \in \mathbb{R}$ we conclude that the problem of null-controllability is equivalent to finding some $u \in L^p([0,T], \mathbb{R}^r)$ which satisfies the vector equation

$$\int_0^T e^{-tA} Bu(t)dt = y_0. \tag{0.6}$$

The second of the two problems mentioned above is the

Problem of Null-Reachability: Given $T > 0$ and $y_T \in \mathbb{R}^n$, find $u \in L^p([0,T], \mathbb{R}^r)$ such that the unique (absolutely continuous) solution $y = y(t)$ of (0.1) and (0.2) for $y_0 = \theta_n$ satisfies the end condition

$$y(T) = y_T. \tag{0.7}$$

Again, on using (0.3) and the non-singularity of e^{tA} for every $t \in \mathbb{R}$ we conclude that the problem of null-reachability is equivalent to finding some $u \in L^p([0,T], \mathbb{R}^r)$ such that the vector equation

$$\int_0^T e^{-tA} Bu(t)dt = - e^{-tA} y_T \tag{0.8}$$

is satisfied.

If we define $Z(t) = e^{-tA}B$ and $c = y_0$ or $c = -e^{-TA}y_T$, then (0.6) and (0.8) can be rewritten in the form

$$\int_0^T Z(t)u(t)dt = c \tag{0.9}$$

where $Z = Z(t)$ is a continuous $n \times r$-matrix function on $\mathbb{R}$ and $c \in \mathbb{R}^n$ is a given vector.

This shows that the problem of null-controllability and the problem of null-reachability are both solvable, if, for every $c \in \mathbb{R}^n$, some $u \in L^p([0,T], \mathbb{R}^r)$ can be found which satisfies (0.9).

The solvability of (0.9) for some given fixed $c \in \mathbb{R}^n$ is a typical <u>finite moment problem</u>. Normally (0.9) is written in the form

$$\int_0^T < Z_j(t),u(t)> dt = c_j$$
$$\text{for } j = 1,\ldots,n \tag{0.10}$$

where $c_1,\ldots,c_n \in \mathbb{R}$ are given "moments", $Z_1(t),\ldots,Z_n(t)$ are the n row vectors of the $n \times r$-matrix $Z(t) = e^{-tA}B$, and $<\cdot,\cdot>$ denotes the scalar product in $\mathbb{R}^r$.

<u>0.2. On the Solvability of the Finite Moment Problem which Contains the Problems of Null-Controllability and Null-Reachability.</u>

At first we give a sufficient condition for the solvability of (0.9) or, equivalently, (0.10) which will also turn out to be necessary. For this purpose we make the following

<u>Definition:</u> The linear system (0.1) is called <u>proper</u> on $[0,T]$, if the row vector functions $Z_j = Z_j(t)$, $j = 1,\ldots,n$, $t \in [0,T]$ of the $n \times r$-matrix function $Z(t) = e^{-tA}B$ are linearly independent on $[0,T]$.

With this definition we can formulate

Theorem 0.1: Let the system (0.1) be proper on [0,T]. Then, for every $c \in \mathbb{R}^n$, there is some $u \in C([0,T], \mathbb{R}^r)$ which satisfies (0.9) or, equivalently, (0.10).

Proof: If we define, for any given $y \in \mathbb{R}^n$,

$$u(t) = u_y(t) = Z(t)^T y = \sum_{j=1}^{n} y_j z_j(t)^T, \quad t \in [0,T], \tag{0.11}$$

then $u = u_y$ is in $C([0,T], \mathbb{R}^r)$ and satisfies (0.9) or, equivalently, (0.10), if and only if $y \in \mathbb{R}^r$ is a solution of the linear system

$$\int_0^T Z(t) Z(t)^T \, dt \, y = c. \tag{0.12}$$

Due to the properness of the linear system (0.1) on [0,T] the $n \times n$-matrix $\int_0^T Z(t) Z(t)^T \, dt$ is (symmetric and) positive definite, hence non-singular. Therefore the linear system (0.12) has exactly one solution $y \in \mathbb{R}^n$ and by the definition (0.11) we obtain a solution $u = u_y$ of (0.9) which is in $C([0,T], \mathbb{R}^r)$. This completes the proof.

Conversely we also can prove

Theorem 0.2: If, for every $c \in \mathbb{R}^n$, there is some $u \in L^\infty([0,T], \mathbb{R}^r)$ ($\subseteq L^p([0,T], \mathbb{R}^r)$ for $p \in [2, \infty)$) which satisfies (0.9) or, equivalently, (0.10), then the linear system (0.1) is proper on [0,T].

Proof: If we define

$$S(u) = \int_0^T Z(t) u(t) \, dt, \quad u \in L^\infty([0,T], \mathbb{R}^r), \tag{0.13}$$

then we obtain a linear map from $L^\infty([0,T], \mathbb{R}^r)$ into $\mathbb{R}^n$ which, by assumption, is surjective. Let $\| \cdot \|_2$ the Euclidean norm in $\mathbb{R}^r$ and $\mathbb{R}^n$. If we equip the space $L^\infty([0,T], \mathbb{R}^r)$ with the norm

$$\|u\|_\infty = \operatorname*{ess\,sup}_{t \in [0,T]} \|u(t)\|_2, \quad u \in L^\infty([0,T], \mathbb{R}^r),$$

then we conclude that

$$\|S(u)\|_2 = \sup \{y^T S(u) \mid y \in \mathbb{R}^n, \ \|y\|_2 = 1\}$$

$$= \sup \{\int_0^T y^T Z(t) u(t) dt \mid y \in \mathbb{R}^n, \ \|y\|_2 = 1\}$$

$$\leq \int_0^T \|Z(t) u(t)\|_2 \, dt = \int_0^T (\sum_{j=1}^n Z_{jk}(t) u_k(t))^2)^{1/2} \, dt$$

$$\leq \int_0^T (\sum_{j=1}^n (\|Z_j(t)\|_2^2 \cdot \|u(t)\|_2^2)^{1/2} \, dt$$

$$\leq \int_0^T (\sum_{j=1}^n (\|Z_j(t)\|_2^2)^{1/2} \, dt \ \|u\|_\infty.$$

This shows that the linear map $S : L^\infty([0,T], \mathbb{R}^r) \to \mathbb{R}^n$ is continuous. Since $L^\infty([0,T], \mathbb{R}^r)$ and $\mathbb{R}^n$ are Banach spaces ($\mathbb{R}^n$ with the Euclidean norm is even a Hilbert space) and S is surjective, by a well known theorem on continuous linear operators on Banach spaces (see Theorem 1.3.5) the adjoint operator $S^* : \mathbb{R}^n \to L^\infty([0,T], \mathbb{R}^r)^*$ has a bounded inverse. Since S^* is defined by

$$S^*(y)(t) = Z(t)^T y, \quad t \in [0,T], \ y \in \mathbb{R}^n,$$

and maps $\mathbb{R}^n$ into $L^1([0,T], \mathbb{R}^r) \subseteq L^\infty([0,T], \mathbb{R}^r)^*$, we therefore conclude the existence of some constant $m > 0$ such that

$$\|y\|_2 \leq m \int_0^T \|Z(t)^T y\|_2 \, dt = m \int_0^T \|\sum_{j=1}^n y_j Z_j(t)^T\|_2 \, dt$$

for all $y = (y_1, \ldots, y_n)^T \in \mathbb{R}^n.$

This in turn implies that the row vector functions $Z_j = Z_j(t)$, $j = 1, \ldots, n$, $t \in [0,T]$, are linearly independent which finishes the proof.

The properness of the system (0.1) on $[0,T]$ can be shown to be equivalent to the so called <u>Kalman Condition</u>

$$\text{rank}(B \mid AB \mid \ldots \mid A^{n-1}B) = n \tag{0.14}$$

where $(B \mid AB \mid \ldots \mid A^{n-1}B)$ denotes the $n \times (n \cdot r)$-matrix consisting of all columns of the $n \times r$-matrices $B, AB, \ldots, A^{n-1}B$ (for the proof see, for instance, [5]).

Since the Kalman Condition is independent of T, it is equivalent to the properness of the system (0.1) on $[0,T]$ for all $T > 0$.

0.3. On Norm-Bounded Null-Controllability and Null-Reachability.

In this section we assume that the linear system (0.1) is proper on [0,T] such that null-controllability and null-reachability are guaranteed by Theorem 0.1. For physical reasons, however, it is often unrealistic to allow the control function to vary in the whole space $L^p([0,T], \mathbb{R}^r)$. In many cases it is reasonable to require the control to be bounded with respect to a suitable norm in $L^p([0,T], \mathbb{R}^r)$. In order to define such a norm we choose the Euclidean norm $\|\cdot\|_2$ in $\mathbb{R}^r$ and define, for every $u \in L^p([0,T], \mathbb{R}^r)$ and every $p \in [2,\infty]$,

$$\|u\|_{p,T} = \begin{cases} \left(\int_0^T \|u(t)\|_2^p \, dt \right)^{\frac{1}{p}} & \text{for } 2 \le p < \infty, \\[2em] \operatorname*{ess\,sup}_{t \in [0,T]} \|u(t)\|_2 & \text{for } p = \infty. \end{cases}$$

Now we can prove the

__Theorem 0.3:__ Let $T > 0$, $c \in \mathbb{R}^n$, and $M > 0$ be given.

Then the following two assertions hold true:

a) If there is a solution $u \in L^p([0,T], \mathbb{R}^r)$ of (0.9) or, equivalently, (0.10) with $\|u\|_{p,T} \le M$, then it follows that

$$c^T y \le M \left(\int_0^T \|Z(t)^T y\|_2^q \, dt \right)^{\frac{1}{q}} \quad \text{for all } y \in \mathbb{R}^n \tag{0.15}$$

where $\frac{1}{p} + \frac{1}{q} = 1$ for $p \in [2,\infty)$ and $q = 1$ for $p = \infty$.

b) Conversely, if (0.15) holds true, then there is some solution $u \in L^p([0,T], \mathbb{R}^r)$ of (0.9) or, equivalently, (0.10) such that $\|u\|_{p,T} \le M$.

__Proof:__ a) Let $y \in \mathbb{R}^n$ be chosen arbitrarily. Then it follows that

$$c^T y = \int_0^T \langle u(t), Z(t)^T y \rangle \, dt \le \int_0^T \|u(t)\|_2 \, \|Z(t)^T y\|_2 \, dt$$

$$\leq \begin{cases} (\int_0^T \|u(t)\|_2^p \, dt)^{\frac{1}{p}} (\int_0^T \|z(t)^T y\|_2^q \, dt)^{\frac{1}{q}} & \text{for } p \in [2,\infty), \\[2em] \operatorname*{ess\,sup}_{t \in [0,T]} \|u(t)\|_2 \int_0^T \|z(t)^T y\|_2 \, dt & \text{for } p = \infty \end{cases}$$

$$\leq M (\int_0^T \|z(t)^T y\|_2^q)^{\frac{1}{q}} \quad \text{for } p \in [2,\infty]$$

which proves (0.15).

b) Let (0.15) be satisfied. Then we consider the n-dimensional subspace V of $L^q([0,T], \mathbb{R}^r)$ which is spanned by $z_1^T, \ldots, z_n^T$. For every $v \in V$ we define a linear functional $\ell : V \to \mathbb{R}$ by

$$\ell(v) = c^T y \quad \text{where} \quad v(t) = \sum_{j=1}^n y_j z_j(t)^T = z(t)^T y. \tag{0.16}$$

Since $z_1^T, \ldots, z_n^T \in V$ are linearly independent (due to the assumed properness of (0.1) on [0,T]), this linear functional is well defined. Moreover, ℓ is bounded on V with norm

$$\|\ell\| = \sup\{\ell(v) \mid v \in V, \ (\int_0^T \|v(t)\|_2^q \, dt)^{\frac{1}{q}} = 1\} \leq M$$

because of (0.15). By the theorem of Hahn-Banach ℓ can be extended to a bounded linear functional ℓ on all of $L^q([0,T], \mathbb{R}^r)$ which has the same norm and which can be represented as

$$\ell(v) = \int_0^T \langle u(t), v(t) \rangle \, dt, \quad v \in L^q([0,T], \mathbb{R}^r), \tag{0.17}$$

for some $u \in L^p([0,T], \mathbb{R}^r)$ with $\|u\|_{p,T} \leq M$. If we choose particularly $v = z_j$ for $j = 1, \ldots, n$ we infer from (0.16), (0.17) that (0.10) is satisfied which finishes the proof.

If we particularly choose $c = y_0$ (see (0.6)) or $c = - e^{-TA} y_T$ (see (0.8)), then (0.15) is a necessary and sufficient condition for the problem of norm-bounded null-controllability or null-reachability ot be solvable.

The question now arises how Theorem 0.3 can be used in order to decide, for $T > 0$ and $c \in \mathbb{R}^n$ being given, for which $M > 0$ there exists a solution $u \in L^p([0,T], \mathbb{R}^r)$ of (0.9) such that $\|u\|_{p,T} \leq M$.

Obviously the smallest number $M \geq 0$ for which (0.15) can be satisfied is given by

$$M(T,c) = \sup\{c^T y \mid y \in \mathbb{R}^n, \quad (\int_0^T \|z(t)^T y\|_2^q \, dt)^{\frac{1}{q}} = 1\}. \tag{0.18}$$

Thus, by Theorem 0.3 (b), for every $M \geq M(T,c)$ there exists some solution $u \in L^p([0,T], \mathbb{R}^r)$ of (0.9) with $\|u\|_{p,T} \leq M$. On the other hand, if $u \in L^p([0,T], \mathbb{R}^r)$ is any solution of (0.9), then, for every $y \in \mathbb{R}^n$, it follows that

$$c^T y = \int_0^T u(t)^T z(t) y \, dt \leq \int_0^T \|u(t)\|_2 \, \|z(t)^T y\|_2 \, dt$$

$$\leq \|u\|_{p,T} \, (\int_0^T \|z(t)^T y\|_2^q \, dt)^{\frac{1}{q}}.$$

This, in connection with (0.18) and Theorem 0.3 b), implies that

$$M(T,c) = \min\{\|u\|_{p,T} \mid u \in L^p([0,T], \mathbb{R}^r) \text{ and satisfies } (0.9)\}. \tag{0.19}$$

Thus $M(T,c)$ defined by (0.18) is also the smallest possible norm of a solution of (0.9) in $L^p([0,T], \mathbb{R}^r)$.

Due to the properness of the linear system (0.1) on $[0,T]$ the set

$$W_T = \{y \in \mathbb{R}^n \mid (\int_0^T \|z(t)^T y\|_2^q \, dt)^{\frac{1}{q}} = 1\} \tag{0.20}$$

is compact which implies the existence of some $y_T \in W_T$ such that $c^T y_T = M(T,c)$.

Let

$$U_T = \{u \in L^p([0,T], \mathbb{R}^r) \mid \int_0^T z(t) u(t) \, dt = c\}. \tag{0.21}$$

For every $u_T \in U_T$ with $\|u\|_{p,T} = M(T,c)$ and every $y_T \in W_T$ with $c^T y_T = M(T,c)$ we then infer that

$$\int_0^T u_T(t)^T z(t)^T y_T \, dt = \|u_T\|_{p,T}. \tag{0.22}$$

Because of

$$\int_0^T u_T(t)^T z(t)^T y_T \, dt \leq \int_0^T \|u_T(t)\|_2 \, \|z(t)^T y_T\|_2 \, dt$$

$$\leq \|u_T\|_{p,T} \, \underbrace{(\int_0^T \|z(t)^T y_T\|_2^q \, dt)^{\frac{1}{q}}}_{= 1}$$

the equality (0.22) implies that

$$\int_0^T u_T(t)^T z(t)^T y_T \, dt = \int_0^T \|u_T(t)\|_2 \, \|z(t)^T y_T\|_2 \, dt$$

and in turn

$$u_T(t)^T z(t)^T y_T = \|u_T(t)\|_2 \, \|z(t)^T y_T\|_2 \tag{0.23}$$

$$\text{for almost all } t \in [0,T].$$

This is only possible, if, for almost every $t \in [0,T]$, there exists some $\alpha(t) \in \mathbb{R}$ such that $\alpha(t) \geq 0$ and

$$u_T(t) = \alpha(t) z(t)^T y_T. \tag{0.24}$$

Assumption: For every $y \in \mathbb{R}^n$ with $y \neq \theta_n$ it follows that

$$z(t)^T y \neq \theta_r \quad \text{for almost all } \ t \in [0,T]. \tag{0.25}$$

Under this assumption we conclude from (0.24) that

$$\alpha(t) = \frac{\|u_T(t)\|_2}{\|z(t)^T y_T\|_2} \quad \text{for almost all } t \in [0,T],$$

hence,

$$u_T(t) = \frac{\|u_T(t)\|_2}{\|z(t)^T y_T\|_2} \, z(t)^T y_T \quad \text{for almost all } \ t \in [0,T].$$

Insertion into (0.22) leads to

$$\int_0^T \|u_T(t)\|_2 \, \|z(t)^T y_T\|_2 \, dt = \|u_T\|_{p,T} \, (\int_0^T \|z(t)^T y_T\|_2^q \, dt)^{\frac{1}{q}} \ .$$

Let $p = \infty$ $(\Longrightarrow q = 1)$ then

$$\int_0^T \underbrace{(\|u_T(t)\|_2 - \operatorname*{ess\,sup}_{s \in [0,T]} \|u_T(s)\|_2)}_{\leq 0} \|z(t)^T y_T\|_2 \, dt = 0$$

which implies

$$\|u_T(t)\|_2 = \operatorname*{ess\,sup}_{s \in [0,T]} \|u_T(s)\|_2 = \|u_T\|_{\infty,T} = M(T,c) = c^T y_T$$

and in turn

$$u_T(t) = \frac{c^T y_T}{\|z(t)^T y_T\|_2} z(t)^T y_T \quad \text{for almost all} \quad t \in [0,T].$$

The next special case is $p = q = 2$. This implies

$$\int_0^T \|u_T(t)\|_2 \, \|z(t)^T y_T\|_2 \, dt = \|u_T\|_{2,T} \cdot \|z(\cdot)^T y_T\|_{2,T}$$

which is only possible, if

$$\|u_T(t)\|_2 = \alpha \|z(t)^T y_T\|_2 \quad \text{for all} \quad t \in [0,T]$$

and some $\alpha \geq 0$. From this we deduce that

$$\|u_T\|_{2,T} = \alpha = c^T y_T$$

and therefore

$$u_T(t) = \alpha z(t)^T y_T = (c^T y_T) z(t)^T y_T \quad \text{for all} \quad t \in [0,T].$$

If we define $\hat{y}_T = (c^T y_T) y_T$, then it follows that

$$\int_0^T z(t) z(t)^T \, dt \, \hat{y}_T = c$$

which uniquely determines $\hat{y}_T \in \mathbb{R}^n$ and leads to

$$u_T(t) = z(t)^T \hat{y}_T, \quad t \in [0,T],$$

as unique solution to the minimum norm problem (0.19) for $p = 2$.

Finally, we consider the case $p \in (2,\infty)$ ($\implies q \in (1,2)$). Then we have the equality

$$\int_0^T \|u_T(t)\|_2 \, \|z(t)^T y_T\|_2 \, dt = (\int_0^T \|u_T(t)\|_2^p \, dt)^{\frac{1}{p}} (\int_0^T \|z(t)^T y_T\|_2^q \, dt)^{\frac{1}{q}}$$

which is only possible, if

$$\|u_T(t)\|_2^p = \alpha \ \|z(t)^T y_T\|_2^q \quad \text{for all} \quad t \in [0,T]$$

and some $\alpha \geq 0$. From this we deduce that

$$\| u_T\|_{p,T} = \alpha^{\frac{1}{p}} \ \underbrace{(\int_0^T \| z(t)^T y_T\|_2^q \ dt)}_{= 1}^{1-\frac{1}{q}} = \alpha^{\frac{1}{p}}$$

and hence

$$\begin{aligned}
u_T(t) &= \frac{\| u_T(t)\|_2}{\|z(t)^T y_T\|_2} \ z(t)^T y_T \\[2mm]
&= \alpha^{\frac{1}{p}} \ \|z(t)^T y_T\|_2^{q-2} \ z(t)^T y_T \\[2mm]
&= (c^T y_T) \ \|z(t)^T y_T\|_2^{q-2} \ z(t)^T y_T
\end{aligned}$$

for almost all $t \in [0,T]$.

This formula also holds true for the two special cases $p = \infty$, $q = 1$, and $p = q = 2$.

0.4. On Time-Minimal Null-Controllability and Null-Reachability.

In the following we assume the right-hand side of (0.9) to depend on T, i.e., $c = c(t)$ (as in the case of null-reachability where $c(T) = - e^{-TA} y_T$). Let us further assume that $c : \mathbb{R} \to \mathbb{R}^n$ is continuous. For every $p \in [2,\infty]$ we define $X = L^p([0,\infty], \mathbb{R}^n)$ and put, for every $T > 0$,

$$U_{p,T}(M) = \{u \in X \mid \ \|u\|_{p,T} \ \leq M\} \tag{0.26}$$

where

$$\|u\|_{p,T} = \begin{cases} (\int_0^T \|u(t)\|_2^p \ dt)^{\frac{1}{p}} & \text{for} \quad 2 \leq p < \infty, \\[4mm] \underset{t \in [0,T]}{\text{ess sup}} \ \|u(t)\|_2 & \text{for} \quad p = \infty \ . \end{cases} \tag{0.27}$$

<u>Assumption:</u> There exists some $\hat{T} > 0$ and some $\hat{u} \in U_{p,\hat{T}}(M)$ such that

$$\int_0^T Z(t)u(t) \ dt = c(T) \tag{0.28}$$

holds true for $T = \hat{T}$ and $u = \hat{u}$.

Then the <u>minimum time</u>

$$T(M) = \inf\{T > 0 \mid \text{There is some } u \in U_{p,T}(M) \tag{0.29}$$

$$\text{which satisfies (0.28)}\}$$

is well-defined and the following existence theorem can be proved.

<u>Theorem 0.4:</u>   Under the above assumption there exists some $u_M \in U_{p,T(M)}(M)$ such that

$$\int_0^{T(M)} Z(t) u_M(t) \, dt = c(T(M)) \tag{0.30}$$

is satisfied.

The proof is the same as that of Theorem 1.5.1 and will therefore be omitted. Theorem 0.4 is immediately applicable to the problem of norm-bounded null-controllability or null-reachability and states that, if this is solvable for some $T = \hat{T} > 0$, then it is also solvable for the minimal time $T = T(M)$ defined by (0.29) where $c(T) = y_0$ or $c(T) = - e^{-TA} y_T$ for all $t > 0$ ($y_T \in R^n$ being the same vector for all $T > 0$).

Every solution $u_M \in U_{p,T(M)}$ of (0.30) is called <u>time-minimal</u>. The next step is to derive necessary and sufficient conditions for time-minimal solutions. The above assumption is satisfied, if we choose $T > 0$ arbitrarily and $M > 0$ such that $M \geq M(\hat{T}, c(\hat{T}))$ where $M(T,c)$ is given by (0.19). As a consequence of (0.19) and Theorem 0.4 we then obtain the inequality

$$M(T(M), c(T(M))) \leq M \tag{0.31}$$

and the question arises under which condition we do have equality in (0.31).

From now on we assume that

$$c(T) = c \neq \theta_n \quad \text{for all } T \in R. \tag{0.32}$$

Then, as an immediate consequence of (0.19), we obtain the implication

$$0 < T_1 < T_2 \implies M(T_1, c) \geq M(T_2, c). \tag{0.33}$$

So far we have always tacitly assumed the linear system (0.1) to be proper on $[0,T]$ for every $T > 0$ (which is equivalent to the Kalman Condition (0.14) as mentioned above).

This can be used in order to prove the left continuity of the function $T \to M(T,c)$ for $c \in \mathbb{R}^n$, $c \neq \Theta_n$ being fixed. For this purpose we define the function

$$\nu(T,c) = \inf\{\|Z(\cdot)^T y\|_{q,T} \mid y \in \mathbb{R}^n \text{ with } c^T y = 1\} \tag{0.34}$$

where $\frac{1}{p} + \frac{1}{q} = 1$ for $p \in [2,\infty)$ and $q = 1$ for $p = \infty$.

Then one can prove the following

<u>Lemma 0.5:</u> For every $c \in \mathbb{R}^n$ with $c \neq \Theta_n$ it follows that $M(T,c) > 0$, $\nu(T,c) = \frac{1}{M(T,c)}$, and the infimum in (0.34) is attained.

The proof of this lemma is left as an exercise. As a consequence of it we obtain the

<u>Theorem 0.6:</u> For every $c \in \mathbb{R}^n$ with $c \neq \Theta_n$ the function $T \to M(T,c)$, $T \in (0,\infty)$, is left continuous, i.e., for every $T^* > 0$ we have

$$\lim_{T \to T^* - 0} M(T,c) = M(T^*,c).$$

<u>Proof:</u> By virtue of Lemma 0.5 it suffices to show that the function $T \to \nu(T,c)$ (0.34) is left continuous. Therefore, let $T^* > 0$ be choosen and choose $T_0 \in (0,T^*)$ arbitrarily. For every $T \in [T_0,T^*]$ we select $y_T \in \mathbb{R}^n$ with $c^T y_T = 1$ such that $\|Z(\cdot)^T y_T\|_{q,T} = \nu(T,c)$ which is possible due to Lemma 0.5.

Then we obtain from

$$0 \le \nu(T^*,c)^q - \nu(T,c)^q = \int_0^{T^*} \|Z(t)^T y_*\|_2^q \, dt - \int_0^T \|Z(t)^T y_T\|_2^q \, dt$$

$$\le \int_0^{T^*} \|Z(t)^T y_T\|_2^q \, dt - \int_0^T \|Z(t)^T y_T\|_2^q \, dt = \int_T^{T^*} \|Z(t) y_T\|_2^q \, dt \tag{0.35}$$

$$\le \int_T^{T^*} \|Z(t)\|_2^q \, dt \, \|y_T\|_2^q$$

where

$$\|z(t)\|_2 = (\sum_{j=1}^{n} \sum_{k=1}^{r} z_{jk}(t)^2)^{1/2} .$$

In order to estimate $\|y_T\|_2$ we make use of the fact that due to the properness of (0.1) on $[0,T_0]$ it follows that

$$m_{T_0} = \min \{ (\int_0^{T_0} \|z(t)^T y\|_2^q \, dt)^{\frac{1}{q}} \mid y \in \mathbb{R}^n, \|y\|_2 = 1 \} > 0.$$

This implies

$$m_{T_0} \|y\|_2 \le (\int_0^{T_0} \|z(t)^T y\|_2^q \, dt)^{\frac{1}{q}} \quad \text{for all} \quad y \in \mathbb{R}^n.$$

In particular we obtain

$$m_{T_0} \|y_T\|_2 \le (\int_0^{T_0} \|z(t)^T y_T\|_2^q \, dt)^{\frac{1}{q}} \le (\int_0^{T} \|z(t)^T y_T\|^q dt)^{\frac{1}{q}}$$

$$= \nu(T,c) \le \nu(T^*,c)$$

Insertion into (0.35) leads to

$$0 \le \nu(T^*,c)^q - \nu(T,c)^q \le \int_T^{T^*} \|z(t)\|_2^q \, dt \, (\frac{\nu(T^*,c)}{m_{T_0}})^q$$

which implies

$$\lim_{T \to T^*-0} \nu(T,c) = \nu(T^*,c)$$

and completes the proof of Theorem 0.6.

This result is crucial for

Theorem 0.7: Let the Kalman Condition (0.14) be satisfied. Further we assume $M > 0$ be given such that there is some $\hat{T} > 0$ with $M \ge M(\hat{T},c)$ (0.19). Then the following implication holds true

$$T = T(M) \implies M(T,c) = M \tag{0.36}$$

where $T(M)$ is the minimum time defined by (0.29) for $c(T) = c \ne \theta_n$.

Proof: By (0.31) we have the inequality $M(T(M),c) \le M$. In order to prove equality we choose a sequence $(T_k)_{k \in \mathbb{N}}$ in $(0,T(M))$ (from Theorem 0.4 and $c \ne \theta_n$ it follows that $T(M) > 0$) with $\lim_{k \to \infty} T_k = T(M)$. For every

$k \in \mathbb{N}$ it necessarily follows that $M(T_k,c) \geq M$ for otherwise we would have $T_k \in (0,T(M))$ and some $u_k \in U_{p,T_k}(M)$ with $\int_0^{T_k} Z(t)u_k(t)\, dt = c$ contradicting the minimality of $T(M)$. By Theorem 0.6 we therefore conclude that $M(T(M),c) = \lim_{k \to \infty} M(T_k,c) \geq M$ which implies $M(T(M),c) = M$ and proves the implication (0.36).

This implication gives a necessary condition for some $T > 0$ to be the minimum time being defined by (0.29) and it also implies that every time minimal solution of $\int_{T(M)}^{T} Z(t)u(t)\, dt = c$ with $\|u\|_{p,T} \leq M$ is a minimum norm solution of $\int_0^{T} Z(t)u(t)\, dt = c$.

More important than (0.36) is the opposite implication

$$M(T,c) = M \implies T = T(M) \tag{0.37}$$

which yields a sufficient condition for some time $T > 0$ to be the minimum time $T(M)$ defined by (0.29). Obviously (0.37) holds true, if $M(T(M),c) = M$ is satisfied and if the function $T \to M(T,c)$ is strictly decreasing (by (0.35) we know already that this function is non increasing). A condition which guarantees both is the

Normality Condition: For every $T > 0$ and every $y \in \mathbb{R}^n$ with $y \neq \theta_n$ the components of the vector function $Z(\cdot)^T y$ only vanish on a subset of $[0,T]$ of (Lebesgue-) measure zero.

This condition implies the properness of the linear system (0.1) on $[0,T]$ for every $T > 0$ and therefore guarantees the equation

$$M(T(M),c) = M,$$

if there is some $\hat{T} > 0$ such that $M(\hat{T},c) \leq M$ (see Theorem 0.7). In [5] it is shown that the normality condition is equivalent to the statement that every $n \times n$-matrix of the form

$$(B_i \mid AB_i \mid \ldots \mid A^{n-1}B_i), \quad i = 1,\ldots,r,$$

with $B_i = i$-th column vector of B is non-singular. This condition implies the Kalman Condition (0.14).

Lemma 0.8: If the normality condition holds, then the function $T \to M(T,c)$ is strictly decreasing.

Proof: By Lemma 0.5 it suffices to show that the function
$T \to \nu(T,c)$ (0.34) is strictly increasing. Thus, let T_1, $T_2 \in \mathbb{R}$ be
given such that $0 \le T_1 < T_2$. By Lemma 0.5 there exist $y_{T_i} \in \mathbb{R}^n$ with
$c^T y_{T_i} = 1$ for $i = 1,2$ such that $c^T y_{T_i} = \nu(T_i,c)$ for $i = 1,2$. This
implies

$$\nu(T_1,c)^q - \nu(T_2,c)^q = \int_0^{T_1} \|z(t)^T y_{T_1}\|_2^q \, dt - \int_0^{T_2} \|z(t)^T y_{T_2}\|_2^q \, dt$$

$$\le \int_0^{T_1} \|z(t)^T y_{T_2}\|_2^q \, dt - \int_0^{T_2} \|z(t)^T y_{T_2}\|_2^q \, dt = - \int_{T_1}^{T_2} \|z(t)^T y_{T_2}\|_2 \, dt$$

where

$$\int_{T_1}^{T_2} \|z(t)^T y_{T_2}\|_2 \, dt > 0,$$

since otherwise $z(t)^T y_{T_2} = \theta_r$ for $t \in [T_1,T_2]$ which, by the normality
condition implies $y_{T_2} = \theta_n$ and contradicts $c^T y_{T_2} = 1$. Therefore
$\nu(T_1,c)^q < \nu(T_2,c)^q$ and hence $\nu(T_1,c) < \nu(T_2,c)$ which completes the
proof.

Summarizing we obtain the

Theorem 0.9: Let the normality condition hold and let there exist
some $\hat{T} > 0$ with $M \ge M(\hat{T},c)$ (0.19). Then the following equivalence
holds true:

$$T = T(M) \iff M(T,c) = M. \tag{0.39}$$

Due to the assumption (0.32) this result can only be applied to the
problem of time-minimal null-controllability, since in this case
the right-hand side of (0.9) does not depend on T. The equivalence
(0.39) shows that the problem of finding the minimum time T(M) given
by (0.29) can be replaced by the problem of solving the equation
M(T,c) = M. This can be done by any method for solving nonlinear
equations for a real unknown which only require function evaluations
(like the secant method or inverse quadratic interpolation). In this
case the evaluation of the function $M(\cdot,c)$ requires the solution of
a minimum norm moment problem of the form (0.9).

For a more detailed representation of this method for solving time-
minimal control problems of finite dimension the reader is referred
to [6].

0.5 An Outline of Infinite-Dimensional Linear Control Problems.

The finite-dimensional linear control problem being intoduced in Section 0.1 can be generalized to an abstract linear control problem as follows: Let Y and H be Hilbert spaces over $\mathbb{R}$ (which replace $\mathbb{R}^n$ and $\mathbb{R}^r$ equipped with the Euclidean norm) and let A and B be linear operators defined on a dense subspace D(A) of Y and on H, respectively, and mapping D(A) and H into Y, respectively. Then the linear system (0.1) can be replaced by an abstract linear differential equation of the same form as (0.1), namely,

$$\dot{y}(t) = Ay(t) - Bu(t), \quad t \in \mathbb{R}, \tag{0.1'}$$

where y is a function from $\mathbb{R}$ into D(A) which is differentiable in the sense that, for every $t \in \mathbb{R}$,

$$\lim_{h \to 0} \left\| \frac{1}{h}(y(t+h) - y(t)) - \dot{y}(t) \right\|_Y = 0 \tag{0.40}$$

is satisfied and u is a given function from $\mathbb{R}$ into H. If $B : H \to Y$ is bounded and $u \in C_b(\mathbb{R},H)$ = vector space of bounded continuous functions $u : \mathbb{R} \to H$ and if

$$y(0) = y_0 \tag{0.41}$$

where $y_0 \in D(A)$, then by a classical solution of (0.1'), (0.41) one understands a function $y : \mathbb{R} \to D(A)$ such that the functions $Ay(\cdot)$ and $\dot{y}(\cdot)$ are continuous and (0.1') and (0.41) are satisfied.

If A is the infinitesimal generator of a strongly continuous semigroup $S : \mathbb{R} \to L(Y,Y)$ = vector space of bounded linear operators from Y into Y, if $y_0 \in D(A)$, if $B : H \to Y$ is bounded and $u \in C_b(\mathbb{R},H)$, then there exists exactly one classical solution of (0.1'), (0.41) which is given by

$$y(t) = S(t)y_0 - \int_0^t S(t-s)Bu(s) \, ds, \quad t \in \mathbb{R}, \tag{0.42}$$

where the integral is defined in the sense of Bochner (see, for instance, [3]). This is the generalization of the formula (0.3) of variation of constants in the case of a finite-dimensional linear system (0.1) with initial condition (0.2).

The assumptions $y_0 \in D(A)$ and $u \in C_b(\mathbb{R},H)$ are normally too strong for realistic control problems. If $y_0 \in Y$ and (as it is normally the case) $u \in L^\infty(\mathbb{R},H)$ = vector space of all functions $f : \mathbb{R} \to H$ such

that $\|f(\cdot)\|_H$ is measurable and essentially bounded, then (0.42)
is taken as definition of a mild solution $y : \mathbb{R} \to Y$ of the initial
value problem (0.1'), (0.41) provided A is the infinitesimal generator of a strongly continuous semigroup and $B : H \to Y$ is a bounded
linear operator.

The problem of controllability now consists, for some given $T > 0$
and some $y_T \in Y$, of finding some $u \in L(\mathbb{R}, H)$ such that the corresponding mild solution $y = y(t)$, $t \in \mathbb{R}$ of (0.1'), (0.2') which is given
by (0.42) satisfies the end condition

$$y(T) = y_T. \tag{0.43}$$

If $y_T = \theta_Y$ = zero element of Y and $y_0 = \theta_Y$ this problem is termed
problem of null-controllability and problem of null-reachability,
respectively.

On using (0.42) the end condition (0.43) can be written in the form

$$\int_0^T S(T-t)Bu(s) \, ds = S(T)y_0 - y_T \tag{0.44}$$

and the problem of controllability turns out to be equivalent to
finding some $u \in L^\infty(\mathbb{R}, H)$ which satisfies the linear operator equation
(0.44) where $T > 0$, y_0 and $y_T \in Y$ are given.

This is exactly the abstract framework of the problem of distributed
control of the temperature of a one-dimensional rod which is considered in Section 2.1.1. In this case we have $Y = L^2(0,1)$ and $H = \mathbb{R}$.
The linear operator A in (0.1') is defined by

$$Az(x) = \frac{d}{dx}\left(p(x)\frac{d}{dx}z(x)\right) + q(x)z(x) \tag{0.45}$$

for $x \in (0,1)$ where $p, q \in C^\infty[0,1]$, $p > 0$, and has the domain

$$D(A) = \{ z \in H^2[0,1] | \quad a_0 z(0) + b_0 z'(0) = 0 \quad \text{and}$$
$$a_1 z(1) + b_1 z'(1) = 0 \} \tag{0.46}$$

where $a_0^2 + b_0^2 > 0$ and $a_1^2 + b_1^2 > 0$. Then D(A) is a dense linear subspace of $L^2[0,1]$. The linear operator $B : \mathbb{R} \to Y$ is defined by

$$Bu = - r(\cdot)u, \quad u \in \mathbb{R}, \tag{0.47}$$

where $r \in Y$ is some fixed function. This operator is obviously bounded.

With these definitions the linear parabolic partial differential
equation (2.1.1) can be written in the abstract form (0.1'), if
we assume the partial derivative with respect to t on the left-
hand side of (2.1.1) to be defined by

$$\lim_{h \to 0} \left\| \frac{1}{h} (y(\cdot,t+h) - y(\cdot,t)) - y_t(\cdot,t) \right\|_{L^2[0,1]} = 0$$

for every $t \in \mathbb{R}$.

It is well known that the operator (0.45) with domain (0.46) is the
infinitesimal generator of a strongly continuous semigroup
S : $\mathbb{R} \to$ L(Y,Y) which is even analytic and can be represented in
the form

$$S(t)y = \sum_{k=1}^{\infty} e^{-\lambda_k t} <y,\gamma_k>_Y \varphi_k, \quad y \in Y, \tag{0.48}$$

where $(-\lambda_k)_{k \in \mathbb{N}}$ with $0 \leq \lambda_1 < \lambda_2 < \ldots \to \infty$ is the sequence of
(simple) eigenvalues and $(\varphi_k)_{k \in \mathbb{N}}$ the corresponding sequence of
(orthonormal) eigenfunctions of A. By $<\cdot,\cdot>_Y$ we denote the scalar
product in Y.

On using the Fourier expansions of y_T and r with respect to the
eigenfunctions φ_k, $k \in \mathbb{N}$, of A and under the assumption that

$$<r,\varphi_k>_Y \neq 0 \quad \text{for all} \quad k \in \mathbb{N}$$

one can verify that the operator equation (0.44) is equivalent to
the infinite system (2.1.14) of exponential moment equations.

The problem of boundary control of the temperature of a one-dimensional rod which is considered in Section 2.1.2 can be transferred
into a problem of distributed control which again leads to an infinite system of exponential moment equations as a necessary and
sufficient condition for controllability.

In Section 2.2 exponential moment problems are studied as moment
problems in Banach spaces. The central result concerning the solvability of such problems is Theorem 2.2.1 which contains Theorem 0.3
for p = ∞ (⟹ q = 1) as a special case. The proof technique in both
cases is the same und essentially rests on the theorem of Hahn-
Banach. Further results are the Theorems 2.2.2 and 2.2.3 which are
concerned with the solvability of finite and infinite moment problems by minimum norm solutions. The result (0.22) for p = ∞ can
also be derived from Theorem 2.2.2.

The application of Theorem 2.2.1 to exponential moment problems in
connection with Theorem 2.4.3 leads to the remarkable statement
that the exponential moment problem is solvable for every time
T > 0, if it is solvable for T = ∞ (see Theorem 2.4.6).

As an immediate consequence of Theorem 2.2.3 it follows from the
solvability of an exponential moment problem for some T > 0 (and
hence for all T > 0) that there exists also a minimum norm solution
(see Theorem 2.4.10). It takes, however, a great deal of effort to
prove that every minimum norm solution is a so called "bang-bang"
solution and hence unique (see Theorem 2.4.12).

As in the case of finite moment problems a reduction of time-minimal
(norm-bounded) solutions to least norm solutions is possible in the
sense that every time-minimal solution of an exponential moment problem is also a minimum norm solution on the minimum time interval
whose norm coincides with the given norm bound (see (2.4.43)). This
implies that every time-minimal solution is also a "bang-bang" solution (see Theorem 2.4.13).

The bang-bang property of time-minimal solutions can also be shown
directly (see Section 2.4.4.2).

The problem of distributed control of the vibration of a one-dimensional medium can also be incorporated into an abstract framework
(see [4] and [7]). For this purpose we replace the partial differential equation (1.1.1) by an abstract wave equation of the form

$$\ddot{y}(t) + Ay(t) = f(t), \quad t \in [0,T], \tag{0.49}$$

for any T > 0. The function f(·) on the right hand side is chosen from
a linear subspace U of the vector space $L^2([0,T],H)$ of all (classes
of) functions f from [0,T] into a Hilbert space H for which the func-
tion $t \rightarrow \|f(t)\|_H$, $t \in [0,T]$, is measurable and $(\int_0^T \|f(t)\|_H^2 \, dt)^{1/2} < \infty$. By
A a linear operator is denoted which maps a dense subspace D(A) of H
into H and which is self-adjoint and positiv definite. By a well-known
result from functional analysis there exists a unique self-adjoint li-
near operator B : D(B) → H with $D(A) \subseteq D(B) \subseteq H$, $B(D(A)) \subseteq D(B)$ and $B^2 =$
B ∘ B = A. This operator is called the "square root" of A and is de-
noted by $A^{1/2}$. The domain $E = D(A^{1/2})$ of $A^{1/2}$ is called the energy
space. If we equip E with the scalar product

$$\langle v,w \rangle_E = \langle A^{1/2}v, A^{1/2}w \rangle_H \quad \text{for} \quad v,w \in E$$

(where $<\cdot,\cdot>_H$ is the scalar product in H), then E becomes a Hilbert space with norm $\|v\|_E = <v,v>_E^{1/2}$, $v \in E$.

Corresponding to (1.1.4) we consider initial conditions of the form

$$y(0) = y_0 \quad \text{and} \quad \dot{y}(0) = y_1 \tag{0.50}$$

where $y_0 \in E = D(A^{1/2})$ and $y_1 \in H$ are given. The first derivative $\dot{y}(t)$ is defined, for every $t \in [0,T]$ by

$$\lim_{h \to 0} \left\| \frac{1}{h} (y(t+h) - y(t)) - \dot{y}(t) \right\|_H = 0.$$

In the case of the vibration of a one-dimensional medium we have $H = L^2(0,1)$, $U = \{r \cdot u(\cdot) \mid u \in L^2(0,T)\}$ where $r \in L^2(0,1)$ is given. The operator A is equal to a linear differential operator L of order 2n with respect to x whose coefficients are time-independent and whose domain D_L is given by (1.1.8). If L is self-adjoint and positive definite, then it has a non-decreasing sequence $(\lambda_j)_{j \in N}$ of real positive eigen-values λ_j with $\lim\limits_{j \to \infty} \lambda_j = \infty$ and a complete orthonormal sequence of eigenfunctions $(e_j)_{j \in N}$ in D_L. In this case the energy space is given by

$$E = \{v \in L^2(0,1) \mid \sum_{j=1}^{\infty} \lambda_j <v,e_j>_{L^2(0,1)}^2 < \infty\}$$

and $A^{1/2}$ is defined by

$$A^{1/2}v = \sum_{j=1}^{\infty} \sqrt{\lambda_j} <v,e_j>_{L^2(0,1)} e_j \quad \text{for} \quad v \in E.$$

The initial data $y_0 \in E$, $y_1 \in H$ and the control functions $f \in U$ on the right-hand side of (0.49) do not allow for a "classical solution" of the initial value problem (0.49), (0.50). This would be a function $y \in C^2([0,T],H) \cap C([0,T],D(A))$ which satisfies (0.49) and (0.50).

Therefore we have to replace the notion of a classical solution by a suitable definition of a weak or generalized solution of (0.49), (0.50). For this purpose we at first give another interpretation to (0.49) by introducing the dual space E^* of E and by defining a linear operator $\tilde{A} : E \to E^*$ by

$$\tilde{A}(v)(w) = <A^{1/2}v, A^{1/2}w>_H \quad \text{for all } w \in E$$

and every $v \in E$. Because of

$$|\tilde{A}(v)(w)| = |<v,w>_E| \le \|v\|_E \|w\|_E \quad \text{for all} \quad v,w \in E$$

the operator $\tilde{A}$ is bounded and $\|\tilde{A}(v)\|_{E^*} \le \|v\|_E$ for all $v \in E$.

Moreover $\tilde{A} : E \to E^*$ is an extension of $A:D(A) \to H$. In order to see this we choose $v \in D(A)$. Then, for every $w \in E$, we obtain

$$\tilde{A}(v)(w) = <Av,w>_H. \tag{0.51}$$

Since $A : D(A) \to H$ is positive definite, there is some constant $m > 0$ such that

$$\|w\|_E^2 = <A^{1/2}w, A^{1/2}w>_H = <Aw,w>_H \ge m\|w\|_H^2,$$

hence

$$\|w\|_H \le \frac{1}{\sqrt{m}} \|w\|_E \quad \text{for all} \quad w \in D(A).$$

This is also true for all $w \in E$, since E is the $\|\cdot\|_E$-norm completion of $D(A)$. Thus every element $h \in H$ can be identified with some $v_h^* \in E^*$ by defining

$$v_h^*(w) = <h,w>_H \quad \text{for all} \quad w \in E,$$

since this implies

$$|v_h^*(w)| \le \|h\|_H \|w\|_H \le \frac{\|h\|_H}{\sqrt{m}} \|w\|_E \quad \text{for all} \quad w \in E.$$

In particular, for every $v \in D(A)$, the element Av can be identified with some $v_{Av}^* \in E^*$ which, by (0.50), is equal to $\tilde{A}$. This shows that $\tilde{A}$ is an extension of A.

Now we replace the abstract wave equation (0.49) by

$$\ddot{y}(t)(w) + \tilde{A}(y(t))(w) = <f(t),w>_H \quad \text{for all} \quad w \in E$$
$$\text{and almost all} \quad t \in [0,T]. \tag{0.52}$$

This, of course, requires $\ddot{y}(t)$ to exist almost everywhere in the sense that

$$\lim_{h \to 0} \|\frac{1}{h} (\dot{y}(t+h) - \dot{y}(t)) - \ddot{y}(t)\|_{E^*} = 0 \tag{0.53}$$

where $\dot{y}(\cdot)$ is defined by (0.40). After these preparations we come to the <u>definition of a weak solution of (0.49), (0.50)</u>: A weak solution of (0.49), (0.50) is a function $y : [0,T] \to H$ such that $y \in C([0,T],E) \cap C^1([0,T],H)$,

$$\lim_{t \to 0+} \|y(t) - y_0\|_E = \lim_{t \to 0+} \|\dot{y}(t) - y_1\|_H = 0,$$

the second derivative $\ddot{y}(t)$ exists for almost all $t \in [0,T]$ in the sense that (0.53) is satisfied and the weak abstract wave equation (0.52) is satisfied.

In this way we have also defined a weak solution of the wave equation (1.1.1) and the initial conditions (1.1.3).

In addition to the above assumptions let us also assume that the operator A in (0.49) has an infinite sequence $(\lambda_j)_{j \in \mathbb{N}}$ of eigenvalues $\lambda_j \in \mathbb{R}$ of finite multiplicity such that $0 < \lambda_1 \leq \lambda_2 \leq \ldots \leq \lambda_n \to \infty$ as $n \to \infty$ and a corresponding sequence $(\psi_j)_{j \in \mathbb{N}}$ of orthonormal eigenelements $\psi_j \in D(A)$. Then we define, for every $t \in [0,T]$, a _sine operator_ $S(t) : H \to E$ and a _cosine operator_ $C(t) : H \to H$, respectively, by

$$S(t)v = \sum_{j=1}^{\infty} \frac{1}{\sqrt{\lambda_j}} \sin\sqrt{\lambda_j}\, t \, <v, \psi_j>_H \psi_j \quad \text{for} \quad v \in H \qquad (0.54a)$$

and

$$C(t)v = \sum_{j=1}^{\infty} \cos\sqrt{\lambda_j}\, t \, <v, \psi_j>_H \psi_j \quad \text{for} \quad v \in H, \qquad (0.54b)$$

respectively. With these definitions one can show that there is a unique weak solution of (0.49), (0.50) in the above sense which is given by

$$y(t) = C(t)y_0 + S(t)y_1 + \int_0^t S(t-s)f(s)\,ds, \quad t \in [0,T], \qquad (0.55)$$

and whose first derivative is given by

$$\dot{y}(t) = -AS(t)y_0 + C(t)y_1 + \int_0^t C(t-s)f(s)\,ds, \quad t \in [0,T], \qquad (0.56)$$

where

$$Av = \sum_{j=1}^{\infty} \lambda_j <v, \psi_j>_H \psi_j \quad \text{for} \quad v \in D(A), \qquad (0.57)$$

$$\int_0^t S(t-s)f(s)\,ds = \sum_{j=1}^{\infty} \frac{1}{\sqrt{\lambda_j}} \int_0^t \sin\sqrt{\lambda_j}\,(t-s) <f(s), \psi_j>_H \, ds \, \psi_j \quad \text{and}$$

$$\int_0^t C(t-s)f(s)ds = \sum_{j=1}^\infty \int_0^t \cos\sqrt{\lambda_j}(t-s) <f(s),\varphi_j>_H ds \, \varphi_j. \text{ The } \underline{\text{problem}}$$

of <u>null-controllability</u> is defined as in Section 1.1.1 and turns out to be equivalent to the problem of finding some f∈U (= subspace of $L^2([0,T],H)$ for T > 0 being given) such that

$$\int_0^T S(T-t)f(t) \, dt = -C(T)y_0 - S(T)y_1,$$

(0.58)

$$\int_0^T C(T-t)f(t) \, dt = AS(T)y_0 - C(T)y_1$$

is satisfied.

If U is of the form $\{r \cdot u(\cdot) \mid u \in L^2(0,T)\}$ (as in the case of distributed control of a one-dimensional vibrating medium), then (0.58) becomes equivalent with a <u>trigonometric moment problem</u> of the form (1.1.23) for an unknown function $u \in L^2(0,T)$ provided

$$<r,\varphi_j>_H \neq 0 \quad \text{for all} \quad j \in \mathbb{N}$$

(0.59)

holds true.

The reduction of the boundary control problem to a problem of distributed control described in Section 1.1.3 (compare with Section 2.1.2) also leads to a trigonometric moment problem of the form (1.1.23), if condition (0.59) is satisfied, which, however, has to be complemented by two additional equations, namely (1.1.44) and (1.1.45). The second of these causes considerable difficulties concerning the solvability of the moment problem (see Sections 1.1.4 and 1.2.2).

The trigonometric moment problem is a typical moment problem in a Hilbert space. Therefore problems of this type are studied in Section 1.2.1. They could be incorporated into the theory of moment problems in general Banach spaces which are investigated in Section 2.2.1. It is, however, more economic to go a direct way which makes a consequent use of the Hilbert space structure.

The problem of distributed null-controllability of a one-dimensional medium is also studied as the problem of solvability of (0.58) (see Section 1.4.1) which is considered as an operator equation in suitable Hilbert spaces. Therefore Section 1.3 is devoted to the solvability of linear operator equations in Banach spaces which turns out also to be applicable to the problem of boundary null-controllability of a one dimensional medium (see Section 1.4.2).

In order to cope with the problem of time-minimal null-controllability by norm-bounded controls for one-dimensional vibrating
media the theory of linear operator equations in Banach spaces investigated in Section 1.3 is refined in Section 1.5. In particular
a general existence theorem for time-minimal controls is proved, a
general maximum-principle for minimum norm controls is derived and
a reduction of time-minimal controllability to minimum norm controllability is shown to be possible. These results are not only applicable to vibrations (see Section 1.5.4) but also to heating processes (see Section 2.4.4.1).

References

[1] Akhiezer, N.J.: The Classical Moment Problem. Edinburgh and Lon-
     don: Oliver and Boyd 1965.

[2] Antosiewicz, H.A.: Linear Control Systems. Arch. Rat. Mech.
     Anal. 12 (1963), 313-324.

[3] Curtain, R.F., and Pritchard, A.H.: Infinite Dimensional Linear
     Systems Theory. Lecture Notes in Control and Information Scien-
     ces No.8. Berlin - Heidelberg - New York: Springer-Verlag 1978.

[4] Fattorini, H.O.: The Time-Optimal Problem for Distributed Control
     of Systems Described by the Wave Equation. In: Aziz, A.K., Win-
     gate, J.W., and Balas, M.J. (Eds.): Control Theory of Systems
     Governed by Partial Differential Equations. New York - San Fran-
     cisco - London: Academic Press 1977.

[5] Hermes, H, and LaSalle, J.P.: Functional Analysis and Time Opti-
     mal Control. New York - London: Academic Press 1969.

[6] Krabs, W.: Einführung in die Kontrolltheorie. Wissenschaftliche
     Buchgesellschaft Darmstadt 1978.

[7] Krabs, W.: On Time-Minimal Distributed Control of Systems Gover-
     ned by an Abstract Wave Equation. Appl. Math. Optim. 13 (1985),
     137-149.

[8] Krasovskii, N.N.: On the Theory of Optimum Control. Prikl. Math.
     Mech. 23 (1959), 625-639.

[9] Marzollo, A.: Controllability and Optimization. International
     Centre for Mechanical Sciences, Courses and Lectures No. 17.
     Wien - New York: Springer-Verlag 1969.

1. Optimal Damping of Vibrations.

1.1 One-Dimensional Vibrating Media.

1.1.1 The Problem of Distributed Control.

We consider the motion of a one-dimensional medium (like a string or a beam) of length 1 (in the position of rest) whose displacement $y(x,t)$ from rest at the point $x \in (0,1)$ and time $t>0$ is controlled along the medium such that y satisfies a differential equation of the form

$$y_{tt}(x,t) + Ly(x,t) = r(x)u(t) \qquad (1.1.1)$$

for $x \in (0,1)$ and $t>0$ where L is a linear differential operator of order 2n with respect to x whose coefficients are time-independent, for instance, $Ly(x,t) = - y_{xx}(x,t)$ in the case of a homogeneous string or $Ly(x,t) = y_{xxxx}(x,t)$ in the case of a(n) (Euler-)beam. The domain D_L of L is assumed to consist of all $z \in H^{2n}(0,1)$ that satisfy 2n boundary conditions of the form

$$B_i^0[z] = \sum_{j=0}^{2n-1} \alpha_{ij} z^{(j)}(0) = 0,$$

$$ \qquad (1.1.2)$$

$$B_i^1[z] = \sum_{j=0}^{2n-1} \beta_{ij} z^{(j)}(1) = 0, \quad i=1,\ldots,n,$$

with $\alpha_{ij}, \beta_{ij} \in \mathbb{R}$.

The function $r=r(x)$ in (1.1.1) is assumed to be fixed and to belong to $L^2(0,1)$ and $u=u(t)$ is considered as a variable control function which is real valued, defined on $[0,\infty)$ and belongs to $L^2(0,T)$ for every $T>0$.

In addition to (1.1.1) the deviation y of the medium from the position of rest is assumed to be subject to homogeneous boundary conditions of the form

$$B_i^0[y(\cdot,t)] = B_i^1[y(\cdot,t)] = 0, \quad i=1,\ldots,n, \qquad (1.1.3)$$

for all $t>0$ and initial conditions

$$y(\cdot,0) = y_0, \ y_t(\cdot,0) = y_1 \text{ on } (0,1) \qquad (1.1.4)$$

are prescribed where y_0 and y_1 are given functions taken from suitable function spaces to be specified later.

The goal now consists of transferring the given initial state
(1.1.4) of vibration to rest within a given time interval $[0,T]$
for some $T>0$ by a suitable choice of u. This leads to the

Problem of Null-Controllability: Given some $T>0$, does there
exist a control function $u \in L^2(0,T)$ such that the corresponding
solution $y = y(x,t,u)$, $x \in [0,1]$, $t \in [0,T]$ of the initial-boundary-
value problem (1.1.1), (1.1.3), (1.1.4) satisfies

$$y(\cdot,T,u) = y_t(\cdot,T,u) = 0 \quad \text{a.e. on } (0,1)? \tag{1.1.5}$$

More realistic from the physical point of view is the problem of
restricted null-controllability where in addition to $u \in L^2(0,T)$ it
is required that

$$\|u\|_{L^2(0,T)} = \left(\int_0^T u(t)^2 dt\right)^{1/2} \leq M \tag{1.1.6}$$

for some given constant $M>0$.

Under the assumption that restricted null-controllability is
possible for some $T = T^*>0$ we also consider the

Problem of Time-Minimal Null-Controllability: Let

$$T(M) = \inf\{T \in (0,T^*] \mid (1.1.5) \text{ and } (1.1.6) \atop \text{are satisfied for some } u \in L^2(0,T)\}. \tag{1.1.7}$$

Question: Does there exist a time-minimal null-control, i.e., a
function $u \in L^2(0,T(M))$ which satisfies (1.1.5) and (1.1.6) for
$T = T(M)$?

Before we are able to attack these problems we have to specify
the spaces for the initial functions y_0 and y_1 and to define in
what sense solutions of (1.1.1), (1.1.3), (1.1.4) exist for y_0, y_1
and u being given.

It is reasonable to require that all states $(y(\cdot,t), y_t(\cdot,t))$, $t>0$,
of the vibration have finite energy. For $y_t(\cdot,t)$ this means that
$y_t(\cdot,t) \in L^2(0,1)$ for all $t>0$ so that y_1 has also to be chosen in
$L^2(0,1)$. In order to specify the choice of y_0 in this sense we
assume that L is self-adjoint and positive definite on

$$D_L = \{z \in H^{2n}(0,1) \mid z \text{ satisfies } (1.1.2)\}. \tag{1.1.8}$$

so that it has a non-decreasing sequence $(\lambda_j)_{j \in N}$ of positive eigenvalues with $\lim_{j \to \infty} \lambda_j = \infty$ and a corresponding complete orthonormal sequence of eigenfunctions $(e_j)_{j \in N}$ in D_L.

The so called space of finite energy with respect to L is then defined by

$$E = \{v \in L^2(0,1) \mid \sum_{j=1}^{\infty} \lambda_j <v,e_j>^2_{L^2(\Omega)} < \infty\}. \tag{1.1.9}$$

with $<\cdot,\cdot>_{L^2(\Omega)}$ denoting the scalar product in $L^2(0,1)$. With this definition we assume y_0 in (1.1.4) to belong to E. We introduce a norm in E by defining

$$\|v\|_E = (\sum_{j=1}^{\infty} \lambda_j <v,e_j>^2_{L^2(0,1)})^{1/2}, \quad v \in E. \tag{1.1.10}$$

Let

$$U = \{u:[0,\infty) \to \mathbb{R} \mid u \in L^2(0,T) \text{ for all } T>0\}. \tag{1.1.11}$$

For a given triple $(y_0,y_1,u) \in E \times L^2(0,1) \times U$ we define a generalized solution of (1.1.1), (1.1.3), (1.1.4) in the first place as a function $y : [0,1] \times [0,\infty) \to \mathbb{R}$ such that

$$y \in L^2((0,1) \times (0,T)) \text{ and } y_t \in L^2((0,1) \times (0,T))$$

for all T>0 where y_t is the t-derivative of y in the sense of distributions. Secondly we require

$$y \in C([0,\infty),E), \quad y_t \in C([0,\infty), L^2(0,1))$$

and

$$\lim_{t \to 0+} \|y(\cdot,t) - y_0\|_E = 0, \quad \lim_{t \to 0+} \|y_t(\cdot,t) - y_1\|_{L^2(0,1)} = 0$$

the last requirement specifying the meaning of (1.1.4). In order to define in what sense (1.1.1) and (1.1.3) are to be satisfied we introduce the dual space E^* of E which consists of all linear functionals $v^* : E \to \mathbb{R}$ of the form

$$v^*(v) = \sum_{j=1}^{\infty} <v,e_j>_{L^2(0,1)} v^*(e_j)$$

with

$$\sum_{j=1}^{\infty} \frac{1}{\lambda_j} v^*(e_j)^2 < \infty.$$

The norm in E^* is given by

$$\|v^*\|_{E^*} = (\sum_{j=1}^{\infty} \frac{1}{\lambda_j} v^*(e_j)^2)^{1/2}, \quad v^* \in E^*$$

It is well-known that D_L (1.1.8) is dense in $L^2(0,1)$ and, since $D_L \subseteq E$ (which can be easily seen), it follows that E is also dense in $L^2(0,1)$. Because of

$$\|v\|_{L^2(0,1)} \leq \frac{1}{\sqrt{\lambda_1}} \|v\|_E \quad \text{for all } v \in E \tag{1.1.12}$$

the space E is continuously imbedded in $L^2(0,1)$. If $L^2(0,1)$ is identified with its dual space, it follows that $L^2(0,1)$ is densely imbedded in E^*, the imbedding being the adjoint of the imbedding of E in $L^2(0,1)$. That $L^2(o,1)$ can be identified with a subspace of E^* also follows directly from (1.1.12). For, if $y \in L^2(0,1)$ is given, we define

$$v_y^*(v) = \langle y,v \rangle_{L^2(0,1)} \quad \text{for all } v \in E$$

and conclude by (1.1.12) that

$$|v_y^*(v)| \leq \|y\|_{L^2(0,1)} \|v\|_{L^2(0,1)} \leq \frac{\|y\|_{L^2(0,1)}}{\sqrt{\lambda_1}} \|v\|_E$$

which implies $v_y^* \in E^*$.

Now we define, for every $v \in E$,

$$A(v)(e_j) = \lambda_j \langle v,e_j \rangle_{L^2(0,1)}, \quad j \in \mathbb{N},$$

and $\tag{1.1.13}$

$$A(v)(w) = \sum_{j=1}^{\infty} \langle w,e_j \rangle_{L^2(0,1)} A(v)(e_j), \quad w \in E.$$

Then, because of

$$\sum_{j=1}^{\infty} \frac{1}{\lambda_j} A(v)(e_j)^2 = \sum_{j=1}^{\infty} \lambda_j \langle v,e_j \rangle_{L^2(0,1)}^2 < \infty,$$

it follows, that A defines a linear mapping from E into E^* which is continuous since, by the last equality,

$$|A(v)|_{E^*} = |v|_E \text{ for all } v \in E$$

(so A is even an isometry). Moreover $A : E \to E^*$ is an extension of $L : D_L \to L^2(0,1)$ because of $A(z)(w) = <Lz,w>_{L^2(0,1)}$ for all $z \in D_L$ and $w \in E$.

Instead of (1.1.1), (1.1.3) we now consider

$$y_{tt}(\cdot,t) + A(y(\cdot,t)) = ru(t) \tag{1.1.14}$$

for almost all $t \in (0,\infty)$ as an equation in E^* and require that a generalized solution y of (1.1.1), (1.1.3), (1.1.4), in addition to the above assumption, satisfies $y_{tt}(\cdot,t) \in E^*$ and (1.1.14) for almost all $t \in (0,\infty)$ where $y_{tt}(\cdot,t)$ is defined by

$$\lim_{h \to 0} \left| y_{tt}(\cdot,t) - \frac{y_t(\cdot,t+h) - y_t(\cdot,t)}{h} \right|_{E^*} = 0$$

and $ru(t)$ is identified with an element in E^* as shown above for an arbitrary $y \in L^2(0,1)$.

After these definitions we can prove

Theorem 1.1.1: For every triple $(y_0,y_1,u) \in E \times L^2(0,1) \times U$ with E and U given by (1.1.9) and (1.1.11), respectively, there exists exactly one generalized solution of (1.1.1), (1.1.3), (1.1.4) in the above sense which is given by

$$y(y,t,u) = \sum_{j=1}^{\infty} (a_j \cos\sqrt{\lambda}_j t + \frac{b_j}{\sqrt{\lambda}_j} \sin\sqrt{\lambda}_j t) e_j(x)$$

$$+ \sum_{j=1}^{\infty} \frac{h_j}{\sqrt{\lambda}_j} \int_0^t u(s) \sin\sqrt{\lambda}_j(t-s) ds\, e_j(x) \tag{1.1.15}$$

with

$$a_j = \int_0^1 y_0(x) e_j(x) dx,$$

$$b_j = \int_0^1 y_1(x) e_j(x) dx, \tag{1.1.16}$$

$$h_j = \int_0^1 r(x) e_j(x) dx, \quad j \in \mathbb{N}.$$

Proof: Let $y = y(x,t,u)$ be a generalized solution of (1.1.1), (1.1.3), (1.1.4) in the above sense. Then, for every $t\epsilon[0,\infty)$, $y(\cdot,t,u)\epsilon E\subseteq L^2(0,1)$ and therefore $y(\cdot,t,u)$ has a representation of the form

$$y(x,t,u) = \sum_{j=1}^{\infty} \varphi_j(t)e_j(x)$$

with

$$\varphi_j(t) = \int_0^1 y(x,t,u)e_j(x)dx, \quad j\epsilon \, \mathbb{N},$$

since $(e_j)_{j\epsilon \, \mathbb{N}}$ is a complete orthonormal system in $L^2(0,1)$.

From $y(\cdot,\cdot,u)\epsilon C([0,\infty),E)$ it follows that $\varphi_j = \varphi_j(t)$, $t\epsilon[0,\infty)$, is continuous for every $j\epsilon \, \mathbb{N}$ and $\lim_{t\to 0+}|y(\cdot,z,u) - y_0|_E = 0$ implies that

$$\varphi_j(0) = a_j \quad \text{for every } j\epsilon \, \mathbb{N}. \tag{1.1.17}$$

From $y_t(\cdot,\cdot,u)\epsilon C([0,\infty),L^2(0,1))$ it follows that $\varphi_j' = \varphi_j'(t)$ exists for every $t\epsilon[0,\infty)$ and is continuous for every $j\epsilon \, \mathbb{N}$. $\lim_{t\to 0+}|y_t(\cdot,t,u) - y_1|_{L^2(0,1)} = 0$ implies that

$$\varphi_j'(0) = b_j \quad \text{for every } j\epsilon \, \mathbb{N}. \tag{1.1.18}$$

From

$$\lim_{h\to 0}|y_{tt}(\cdot,t,u) - \frac{y_t(\cdot,t+h,u) - y(\cdot,t,u)}{h}|_{E^*} = 0$$

for almost all $t\epsilon(0,\infty)$ we conclude that $\varphi_j'' = \varphi_j''(t)$ exists for almost all $t\epsilon(0,\infty)$ and

$$y_{tt}(\cdot,t,u)(w) = \sum_{j=1}^{\infty} \varphi_j''(t)<w,e_j>_{L^2(0,1)} \quad \text{for all } w\epsilon E.$$

Hence (1.1.14) implies that

$$\varphi_j''(t) + \lambda_j\varphi_j(t) = h_j u(t) \quad \text{for almost all } t\epsilon(0,\infty)$$
$$\text{for all } j\epsilon \, \mathbb{N}. \tag{1.1.19}$$

It can be shown that the unique solution of (1.1.17), (1.1.18), (1.1.19) is given by

$$\varphi_j(t) = a_j\cos\sqrt{\lambda}_jt + \frac{b_j}{\sqrt{\lambda}_j}\sin\sqrt{\lambda}_jt + \frac{h_j}{\sqrt{\lambda}_j}\int_0^t u(s)\sin\sqrt{\lambda}_j(t-s)ds$$

for all $j\epsilon\ N$ which implies that every generalized solution of
(1.1.1), (1.1.3), (1.1.4) is necessarily of the form (1.1.15),
(1.1.16).

Conversely, $y = y(x,t,u)$ defined by (1.1.15), (1.1.16) can be
verified to be a generalized solution of (1.1.1), (1.1.3),
(1.1.4) which completes the proof.

Moreover y_t is given by

$$y_t(x,t,u) = \sum_{j=1}^{\infty}\sqrt{\lambda}_j(-a_j\sin\sqrt{\lambda}_jt + \frac{b_j}{\sqrt{\lambda}_j}\cos\sqrt{\lambda}_jt)e_j(x)$$

$$\tag{1.1.20}$$

$$+ \sum_{j=1}^{\infty}h_j\int_0^t u(s)\cos\sqrt{\lambda}_j(t-s)ds\ e_j(x)$$

so that (1.1.5) turns out to be equivalent to

$$\frac{h_j}{\sqrt{\lambda}_j}\int_0^T u(t)\sin\sqrt{\lambda}_j(T-t)dt = -a_j\cos\sqrt{\lambda}_jT - \frac{b_j}{\sqrt{\lambda}_j}\sin\sqrt{\lambda}_jT,$$

$$\tag{1.1.21}$$

$$\frac{h_j}{\sqrt{\lambda}_j}\int_0^T u(t)\cos\sqrt{\lambda}_j(T-t)dt = a_j\sin\sqrt{\lambda}_jT - \frac{b_j}{\sqrt{\lambda}_j}\cos\sqrt{\lambda}_jT$$

for all $j\epsilon\ N$.

We assume that

$$h_j \neq 0 \text{ for all } j\epsilon\ N. \tag{1.1.22}$$

Then (1.1.21) is equivalent with

$$\int_0^T u(t)\cos\sqrt{\lambda}_jt\ dt = c_j^1 = -\frac{b_j}{h_j},$$

$$\tag{1.1.23}$$

$$\int_0^T u(t)\sin\sqrt{\lambda}_jt\ dt = c_j^2 = \frac{a_j\sqrt{\lambda}_j}{h_j}$$

for all $j\epsilon\ N$.

As a result the problem of null-controllability and time-minimal
null-controllability can be rephrased in terms of the solvability
of (1.1.23) for all $j\epsilon\ N$ instead of (1.1.5).

The question of finding some $u\epsilon L^2(0,T)$ which solves (1.1.23)
is a typical (trigonometric) moment problem and will be studied
in full generality in Section 1.2.2. Here we confine ourselves to

1.1.2. An Elementary Case (Part 1).

We assume that, for some $L>0$, the system

$$S = \{\cos\sqrt{\lambda}_jt,\ \sin\sqrt{\lambda}_jt|\ j\epsilon\ \mathbb{N},\ t\geq0\} \tag{1.1.24}$$

is L-periodic and orthogonal on $[0,L]$. We shall give two
representative examples for this case later. Then it follows that

$$\int_0^L(\cos\sqrt{\lambda}_jt)^2dt = \int_0^L(\sin\sqrt{\lambda}_jt)^2dt = \frac{L}{2} \tag{1.1.25}$$

We first consider the case $T = L$.

If $u\epsilon L^2(0,T)$ is a solution of (1.1.23) for all $j\epsilon\ \mathbb{N}$, then, by
Bessel's inequality, it follows that

$$\sum_{j=1}^{\infty}(c_j^1)^2 + (c_j^2)^2 \leq \frac{L}{2}|u|^2_{L^2(0,T)}. \tag{1.1.26}$$

Therefore the condition

$$\sum_{j=1}^{\infty}(\lambda_ja_j^2 + b_j^2)/h_j^2 < \infty \tag{1.1.27}$$

is necessary for the existence of a solution $u\epsilon L^2(0,T)$ of (1.1.23)
for all $j\epsilon\ \mathbb{N}$.

In the following we assume (1.1.27) to be satisfied. If we put

$$u(t) = \frac{2}{L}\sum_{j=1}^{\infty}c_j^1\cos\sqrt{\lambda}_jt + c_j^2\sin\sqrt{\lambda}_jt, \tag{1.1.28}$$

then we obtain a solution $u\epsilon L^2(0,T)$ of (1.1.23) for all $j\epsilon\ \mathbb{N}$ and
$T = L$. Moreover,

$$|u|^2_{L^2(0,T)} = \frac{2}{L}\sum_{j=1}^{\infty}(c_j^1)^2 + (c_j^2)^2 \tag{1.1.29}$$

which shows, by virtue of (1.1.26), that u has the smallest L^2-norm
among all solutions of (1.1.23) for all $j\epsilon\ \mathbb{N}$ in $L^2(0,T)$.
Summarizing we have the following result.

Theorem 1.1.2: Let S (1.1.24) be L-periodic and orthogonal on [0,L] for some L > 0. If (1.1.22) and (1.1.27) are satisfied, then, for T = L, there exists some $u \in L^2(0,T)$ such that (1.1.5) is satisfied. Moreover, u is given by (1.1.28) and has the smallest L^2-norm among all control functions in $L^2(0,T)$ for which (1.1.5) is satisfied. In addition u also satisfies the condition $\int_0^L u(t) \, dt = 0$.

Next we consider the case T ≥ L.

Under the assumptions of Theorem 1.1.2 we immediately get a solution of (1.1.23) for all j∈N defining u(t) according to (1.1.28) for t∈[0,L] and putting u(t) = 0 for t∈[L,T]. Besides this crude way of defining a solution of (1.1.23) for all j∈N we can also find a solution with a smaller L^2-norm which tends to zero as T tends to infinity. Thus restricted null-controllability is guaranteed for every constant M > 0 and sufficiently large T. First we prove

Lemma 1.1.3: Let T = N · L + α for some N∈N and α∈[0,L]. Then $u \in L^2(0,T)$ is an L-periodic solution of (1.1.23) for all j∈N and satisfies $\int_0^T u(t) \, dt = 0$, if and only if u is representable in the form

$$u(t) = \begin{cases} \dfrac{1}{N+1} \, w(t) & \text{for } t \in [0,\alpha), \\[2mm] \dfrac{1}{N} \, w(t) & \text{for } t \in [\alpha,L] \end{cases} \tag{1.1.30a}$$

and

$$u(t+L) = u(t) \quad \text{for} \quad t \in [0,T-L] \tag{1.1.30b}$$

where $w \in L^2(0,L)$ is a solution of (1.1.23) for all j∈N and for T = L which satisfies $\int_0^L w(t) \, dt = 0$.

Proof: Let u, $v \in L^2(0,T)$ be two arbitrary L-periodic functions. Then

$$\int_0^T u(t)v(t) \, dt = \int_0^{NL} u(t)v(t)\,dt + \int_{NL}^{NL+\alpha} u(t)v(t) \, dt$$

$$= \int_0^\alpha (N+1)u(t)v(t)\,dt + \int_\alpha^L N\, u(t)v(t)\,dt \tag{1.1.31}$$

$$= \int_0^L w(t)v(t)\,dt$$

with

$$w(t) = \begin{cases} (N+1)u(t) & \text{for } t \in [0,\alpha), \\ N\,u(t) & \text{for } t \in [\alpha,L]. \end{cases} \tag{1.1.32}$$

If $u \in L^2(0,T)$ is an L-periodic solution of (1.1.23) for all $j \in \mathbb{N}$, then $w \in L^2(0,L)$ defined by (1.1.32) is a solution of (1.1.23) for all $j \in \mathbb{N}$ and $T = L$ which satisfies $\int_0^L w(t)dt = 0$ as a consequence of (1.1.31) and u is representable in the form (1.1.30).

Conversely, if $w \in L^2(0,L)$ is a solution of (1.1.23) for all $j \in \mathbb{N}$ and $T = L$ which satisfies $\int_0^L w(t)dt = 0$, then $u \in L^2(0,T)$ defined by (1.1.30) is an L-periodic solution of (1.1.23) for all $j \in \mathbb{N}$ that satisfies $\int_0^T u(t)dt = 0$ which is also a consequence of (1.1.31) and (1.1.32). This completes the proof.

As an immediate consequence of Lemma 1.1.3 and Theorem 1.1.2 we obtain the

Theorem 1.1.4: Let the assumptions of Theorem 1.1.2 hold and let $T = N \cdot L + \alpha$ for some $N \in \mathbb{N}$ and $\alpha \in [0,L]$. Then there exists an L-periodic solution $u \in L^2(0,T)$ of (1.1.23) for all $j \in \mathbb{N}$ that satisfies $\int_0^T u(t)dt = 0$ which is defined by (1.1.30) with

$$w(t) = \frac{2}{L} \sum_{j=1}^{\infty} c_j^1 \cos\sqrt{\lambda_j}\,t + c_j^2 \sin\sqrt{\lambda_j}\,t \tag{1.1.33}$$

for $t \in [0,L]$.

Moreover, the L^2-norm of u can be estimated in the form

$$\|u\|^2_{L^2(0,T)} \le \frac{N+1}{N^2} \frac{2}{L} \sum_{j=1}^{\infty} (c_j^1)^2 + (c_j^2)^2. \tag{1.1.34}$$

The estimation (1.1.34) in fact shows that

$$\lim_{T \to \infty} \|u\|_{L^2(0,T)} = 0.$$

Theorem 1.1.4 can be sharpened, if we assume that the system $S \cup \{1\}$ with S defined by (1.1.24) is complete in $L^2(0,L)$. Let V be the closure of the span of $S \cup \{1\}$ in $L^2(0,T)$.

<u>Theorem 1.1.5:</u> Let the assumptions of Theorem 1.1.2 hold and let
$T = N \cdot L + \alpha$ for some $N \in \mathbb{N}$ and $\alpha \in [0,L]$. If $S \cup \{1\}$ is complete in
$L^2(0,L)$, then the function $u \in L^2(0,T)$ being defined by (1.1.30) with
w given by (1.1.33) is the unique solution of (1.1.23) for all $j \in \mathbb{N}$
in V which also satisfies $\int_0^T u(t)dt = 0$ and has the smallest L^2-norm
of all solutions $u \in L^2(0,T)$ of (1.1.23) for all $j \in \mathbb{N}$ which satisfy
$\int_0^T u(t)dt = 0$.

<u>Proof:</u> Because of the completeness of $S \cup \{1\}$ in L^2 the function u
being defined by (1.1.30) and (1.1.33) can be uniquely represented
in the form

$$u(t) = a_0 + \frac{2}{L} \sum_{j=1}^{\infty} a_j^1 \cos\sqrt{\lambda_j}\,t + a_j^2 \sin\sqrt{\lambda_j}\,t$$

where

$$a_0 = \frac{1}{L} \int_0^L u(t)dt,$$

$$a_j^1 = \frac{L}{2} \int_0^L u(t)\cos\sqrt{\lambda_j}\,t \; dt,$$

$$a_j^2 = \frac{L}{2} \int_0^L u(t)\sin\sqrt{\lambda_j}\,t \; dt, \quad j \in \mathbb{N}.$$

Hence u is in V. Let $v \in V$ be any solution of (1.1.23) for all $j \in \mathbb{N}$
which also satisfies $\int_0^T v(t)dt = 0$. Then it follows that

$$\langle u-v,w \rangle_{L^2(0,T)} = 0 \quad \text{for all} \quad w \in V. \tag{1.1.35}$$

In particular for $w = u - v$ we conclude $\| u-v \|^2_{L^2(0,T)} = 0$ hence $u = v$.
Let $v \in L^2(0,T)$ be any solution of (1.1.23) for all $j \in \mathbb{N}$ which also
satisfies $\int_0^T v(t)dt = 0$. Then again (1.1.35) follows. In particular
we have

$$\langle u-v,u \rangle_{L^2(0,T)} = 0$$

and therefore

$$0 \leq \langle v-u,v-u \rangle_{L^2(0,T)} = \| v \|^2_{L^2(0,T)} - \| u \|^2_{L^2(0,T)}$$

which completes the proof.

Corollary: Under the assumptions of Theorem 1.1.5 there is exactly one solution $u \in L^2(0,T)$ of (1.1.23) for all $j \in \mathbb{N}$ and $\int_0^T u(t)\,dt$ with $T = L$ which is given by (1.1.28). This implies that, for $T \in (0,L)$, (1.1.23) for all $j \in \mathbb{N}$ has a solution in $L^2(0,T)$, if and only if $u \in L^2(0,T)$ being defined by (1.1.28) satisfies

$\quad u(t) = 0 \quad$ for almost all $t \in (T,L]$

in which case u is the only solution of (1.1.23) for all $j \in \mathbb{N}$ in $L^2(0,T)$.

We conclude this subsection with two special cases:

a)   The Vibrating String.

In this case the differential equation (1.1.1) reads

$$y_{tt}(x,t) - y_{xx}(x,t) = r(x)u(t) \tag{1.1.1'}$$

for $x \in (0,1)$ and $t > 0$. We consider boundary conditions (1.1.3) of the form

$$B_1^0[y(\cdot,t)] = y(0,t) = 0,$$
$$B_1^1[y(\cdot,t)] = y(1,t) = 0, \quad t \geq 0. \tag{1.1.3'}$$

So we have $Lz = -z''$ and the domain of L is given by

$$D_L = \{z \in H^2(0,1) \mid z(0) = z(1) = 0\}. \tag{1.1.8'}$$

The operator L is symmetric and positive definite on D_L with eigenvalues

$$\lambda_j = (j\pi)^2, \quad j \in \mathbb{N},$$

and corresponding orthogonal eigenfunctions

$$e_j(x) = \sqrt{2}\,\sin(j\pi)x, \quad x \in [0,1], \quad j \in \mathbb{N}.$$

The System S (1.1.24) is given by

$$S = \{\cos j\pi t, \sin j\pi t \mid t \geq 0, j \in \mathbb{N}\},$$

is 2-periodic and orthogonal on $[0,2]$ (even orthonormal). Further $S \cup \{1\}$ is complete on $[0,2]$ so that Theorems 1.1.2, 1.1.4, 1.1.5 and the Corollary of the last can be applied.

b) The Vibrating (Euler-) Beam.

In this case the differential equation (1.1.1) reads

$$y_{tt}(x,t) + y_{xxxx}(x,t) = r(x)u(t) \qquad (1.1.1'')$$

for $x \epsilon (0,1)$ and $t>0$. We consider boundary conditions (1.1.3) of the form

$$B_1^0[y(\cdot,t)] = y(0,t) = 0, \quad B_1^1[y(\cdot,t)] = y(1,t) = 0,$$

$$B_2^0[y(\cdot,t)] = y_{xx}(0,t)=0, \quad B_2^1[y(\cdot,t)] = y_{xx}(1,t) = 0,$$

for all $t \geq 0$.

So we have $Lz = z^{(4)}$ and the domain of L is given by

$$D_L = \{z \epsilon H^4(0,1) \mid z(0) = z(1) = z''(0) = z''(1) = 0\}. \qquad (1.1.8'')$$

The operator L is self-adjoint and positive definite on D_L with eigenvalues

$$\lambda_j = (j\pi)^4, \quad j \epsilon \, N,$$

and corresponding orthogonal eigenfunctions

$$e_j(x) = \sqrt{2}\sin(j\pi)x, \quad x \epsilon [0,1], \quad j \epsilon \, N.$$

The system S (1.1.24) is given by

$$S = \{\cos(j\pi)^2 t, \, \sin(j\pi)^2 t \mid t \geq 0, \, j \epsilon \, \mathbb{N}\},$$

is $\frac{2}{\pi}$-periodic and orthogonal on $[0,\frac{2}{\pi}]$. The System $S \cup \{1\}$ is, however, not complete on $[0,\frac{2}{\pi}]$ (see Section 1.1.4) so that only Theorems 1.1.2 and 1.1.4 can be applied.

1.1.3 The Problem of Boundary Control.

We again consider the motion of a one-dimensional medium of length 1 as in Section 1.1.1. But now we assume the displacement $y(x,t)$ at the point $x \epsilon (0,1)$ and time $t>0$ to be governed by a homogeneous differential equation of the form

$$y_{tt}(x,t) + Ly(x,t) = 0 \qquad (1.1.36)$$

for $x \epsilon (0,1)$ and $t>0$ where again L is a linear differential operator of order 2n with respect to x whose coefficients are time-independent. The domain of L is given by (1.1.8). The motion governed by (1.1.36) is now assumed to be controlled on the boundary of the vibrating medium in the form

$$B_i^0[y(\cdot,t)] = 0, \quad B_i^1[y(\cdot,t)] = \delta_{ij}v(t)$$

$$\text{for some fixed } j\in\{1,\ldots,n\} \text{ and } i = 1,\ldots,n \tag{1.1.37}$$

for all $t\geq 0$ where B_i^0 and B_i^1 are defined by (1.1.2), δ_{ij} denotes Kronecker's symbol and the control function v is allowed to vary in

$$V = \{v\in H^2[0,\infty) \mid v(0) = v'(0) = 0\} \tag{1.1.38}$$

where $H^2[0,\infty)$ denotes the space of all functions $v : [0,\infty)\to \mathbb{R}$ such that $v\in H^2(0,T)$ for every $T>0$. Again initial conditions of the form (1.1.4) are prescribed. The assumptions on y_0 and y_1 will be specified later.

The problem of null-controllability is slightly different from the one formulated in Section 1.1.1 and reads as follows:

Given some time $T>0$, does there exist a control function $v\in V$ with

$$v(t) = 0 \text{ for all } t\geq T \tag{1.1.39}$$

which transfers the initial state (y_0,y_1) of vibration at $t = 0$ into rest at $t = T$, i.e.,

$$y(\cdot,T,v) = y_t(\cdot,T,v) = 0 \text{ a.e. on } (0,1) \tag{1.1.40}$$

where $y(\cdot,\cdot,v)$ is the solution of (1.1.36), (1.1.37), (1.1.4) which belongs to the triple (y_0,y_1,v)?

We shall see that, for a reasonable concept of such a solution, (1.1.39) guarantees that the medium stays in rest for all $t\geq T$, if (1.1.40) is achieved.

In the case of distributed control this can also be achieved by simply putting $u(t) = 0$ for $t\geq T$ which is an easy consequence of the concept of generalized solution for (1.1.1), (1.1.3), (1.1.4) as being developed in Section 1.1.1.

In order to give a reasonable definition of a solution of (1.1.36), (1.1.37), (1.1.4) we again assume L to be self-adjoint and positive definite on D_L (1.1.8) thus having a non-decreasing sequence $(\lambda_j)_{j\in N}$ of eigenvalues with $\lim\limits_{j\to\infty}\lambda_j = \infty$ and a corresponding complete orthogonal sequence of eigenfunctions $(e_j)_{j\in \mathbb{N}}$ in D_L.

As a consequence of this assumption there exists a unique solution $r \in C^{2n}[0,1]$ of the boundary value problem

$$Lr(x) = 0, \quad 0 < x < 1,$$

$$B_i^0[r] = 0, \quad B_i^1[r] = \delta_{ij}, \quad i = 1, \ldots, n, \tag{1.1.41}$$

where $j \in \{1, \ldots, n\}$ is the same fixed index as in (1.1.37) and δ_{ij} is again Kronecker's symbol. Let $y_0 \in E$ (1.1.9) and $v \in V$ (1.1.38) be given. If $y = y(\cdot, \cdot, v)$ is a solution of (1.1.36), (1.1.37), (1.1.4) in any reasonable sense, then

$$y^*(x,t,v") = y(x,t,v) - r(x)v(t)$$

is a solution of

$$y_{tt}^*(x,t) + Ly^*(x,t) = - r(x)v"(t) \tag{1.1.1*}$$

for almost all $x \in (0,1)$ and $t > 0$

$$B_i^0[y^*(\cdot,t)] = B_i[y^*(\cdot,t)] = 0, \quad i = 1, \ldots, n \tag{1.1.3*}$$

for all $t \geq 0$ and

$$y^*(\cdot,0) = y_0, \quad y_t^*(\cdot,0) = y_1 \quad \text{a.e. on } (0,1). \tag{1.1.4*}$$

Since $v" \in L^2(0,T)$ for all $T > 0$, by Theorem 1.1.1, there exists a unique generalized solution of (1.1.1*), (1.1.3*), (1.1.4*) in the sense of Section 1.1.1 which is given by

$$y^*(x,t,v") = \sum_{j=1}^{\infty} (a_j \cos\sqrt{\bar{\lambda}_j}t + \frac{b_j}{\sqrt{\bar{\lambda}_j}} \sin\sqrt{\bar{\lambda}_j}t)e_j(x)$$

$$- \sum_{j=1}^{\infty} \frac{h_j}{\sqrt{\bar{\lambda}_j}} \int_0^t v"(s)\sin\sqrt{\bar{\lambda}_j}(t-s)ds \; e_j(x) \tag{1.1.15*}$$

with a_j, b_j, h_j, $j \in \mathbb{N}$, defined by (1.1.16). Moreover, $y_t^*(\cdot, \cdot, v")$ is given by

$$y_t^*(x,t,v") = \sum_{j=1}^{\infty} \sqrt{\bar{\lambda}_j}(-a_j\sin\sqrt{\bar{\lambda}_j}t + \frac{b_j}{\sqrt{\bar{\lambda}_j}} \cos\sqrt{\bar{\lambda}_j}t)e_j(x)$$

$$- \sum_{j=1}^{\infty} h_j \int_0^t v"(s)\cos\sqrt{\bar{\lambda}_j}(t-s)ds \; e_j(x). \tag{1.1.20*}$$

We therefore define

$$y(\cdot, \cdot, v) = y^*(\cdot, \cdot, v") + rv \tag{1.1.42}$$

as the generalized solution of (1.1.36), (1.1.37), (1.1.4) for a given triple $(y_0, y_1, v) \in E \times L^2(0,1) \times V$.

By this definition y is uniquely determined, since y^* depends uniquely on (y_0, y_1, v). It is further guaranteed that (1.1.39) and (1.1.40) imply

$$y(\cdot, t, v) = y_t(\cdot, t, v) = 0 \text{ a.e. on } (0,1)$$

for all $t \geq T$, i.e., if rest is achieved and the control is turned off, then the medium stays in rest. It is, however, no more true that $y(\cdot, t, v) \in E$ for every $t > 0$ unless $r \in E$ which is not always the case.

It is now easy to see that (1.1.39) and (1.1.40) are satisfied, if and only if

$$v(T) = v'(T) = 0, \tag{1.1.43}$$
$$v''(t) = 0 \text{ for almost all } t \geq T,$$

and

$$y^*(\cdot, T, v'') = y_t^*(\cdot, T, v'') = 0. \tag{1.1.5*}$$

Since, for every $v \in V$, we have the representation

$$v(t) = \int_0^t (t-s) v''(s) \, ds, \quad t \geq 0,$$

the condition (1.1.43) is equivalent to

$$\int_0^T t v''(t) \, dt = \int_0^T v''(t) \, dt = 0.$$

The condition (1.1.5*) is equivalent to

$$\frac{h_j}{\sqrt{\lambda_j}} \int_0^T v''(t) \sin\sqrt{\lambda_j}(T-t) \, dt = a_j \cos\sqrt{\lambda_j} T + \frac{b_j}{\sqrt{\lambda_j}} \sin\sqrt{\lambda_j} T,$$

$$\tag{1.1.21*}$$

$$\frac{h_j}{\sqrt{\lambda_j}} \int_0^T v''(t) \cos\sqrt{\lambda_j}(T-t) \, dt = - a_j \sin\sqrt{\lambda_j} T + \frac{b_j}{\sqrt{\lambda_j}} \cos\sqrt{\lambda_j} T$$

for all $j \in \mathbb{N}$. If we assume (1.1.22) to hold, then (1.1.21*) is equivalent to

$$\int_0^T v''(t) \cos\sqrt{\lambda_j} t \, dt = - c_j^1 = \frac{b_j}{h_j},$$

$$\int_0^T v''(t) \sin\sqrt{\lambda_j} t \, dt = - c_j^2 = - \frac{a_j \sqrt{\lambda_j}}{h_j}$$

for all $j \in \mathbb{N}$.

Summarizing we obtain a solution of the problem of null-controllability as follows: For a given T>0 a function $u \in L^2(0,T)$ is determined which satisfies the two equations

$$\int_0^T t\, u(t)\, dt = 0,$$

(1.1.44)

$$\int_0^T u(t)\, dt = 0$$

(1.1.45)

and is a solution of (1.1.23).

If this is possible, the function $v \in V$ defined by

$$v(t) = \begin{cases} - \int_0^t (t-s) u(s)\, ds & \text{for } t \in [0,T], \\ 0 & \text{for } t > T \end{cases}$$

is a solution of (1.1.39), (1.1.40), i.e., solves the problem of null-controllability.

Conversely, if this is the case for some $v \in V$, then $u = - v''$ is a solution of (1.1.44), (1.1.45), (1.1.23) in $L^2(0,T)$.

In order to formulate the problem of restricted null-controllability we now require in addition to $v \in V$ and (1.1.39) that

$$\|v''\|_{L^2(0,T)} \leq M$$

(1.1.46)

for some constant M>0.

The problem of time-minimal null-controllability is now formulated in the same way as in Section 1.1.1.

The solvability of (1.1.44), (1.1.45), (1.1.23) in $L^2(0,T)$ is a typical moment problem in a (real) Hilbert space and will be treated in general in Section 1.2.2. Here again we confine ourselves to

1.1.4 An Elementary Case (Part 2).

As in Section 1.1.2 we assume, for some L>0, the system S (1.1.24) to be L-periodic and orthogonal on [0,L] which implies (1.1.25).

We again first consider the case $T = L$.

We assume (1.1.22) and (1.1.27) to hold the latter condition being necessary for a $u \epsilon L^2(0,T)$ to exist which satisfies (1.1.44), (1.1.45), (1.1.23) for all $j \epsilon \mathbb{N}$. This is shown at the beginning of Section 1.1.2 by Bessel's inequality which also implies

$$\int_0^L t^2 dt \geq \frac{1}{L}(\int_0^L t \, dt)^2 + \frac{2}{L} \sum_{j=1}^{\infty} (\int_0^L t \cos\sqrt{\lambda}_j t \, dt)^2$$

$$+ (\int_0^L t \sin\sqrt{\bar{\lambda}}_j t \, dt)^2 \qquad (1.1.47)$$

and in turn

$$\mu := \frac{L^3}{12} - 2L \sum_{j=1}^{\infty} \frac{1}{\lambda_j} \geq 0 \qquad (1.1.48)$$

by virtue of

$$\int_0^L t^2 dt - \frac{1}{L}(\int_0^L t \, dt)^2 = \frac{L^3}{3} - \frac{L^3}{4} = \frac{L^3}{12},$$

$$\int_0^L t \cos\sqrt{\bar{\lambda}}_j t \, dt = 0 \text{ and } \int_0^L t \sin\sqrt{\bar{\lambda}}_j dt = -\frac{L}{\sqrt{\lambda}_j}, \quad j \epsilon \mathbb{N}.$$

We distinguish two cases:

a) $\mu = 0$. Then (1.1.47) is satisfied as an equality which implies that $z(t) = t$, $t \epsilon [0,L]$, is representable as a Fourier series of the form

$$t = \frac{1}{L} \int_0^L s \, ds + \frac{2}{L} \sum_{j=1}^{\infty} \int_0^L s \sin\sqrt{\bar{\lambda}}_j s \, ds \sin\sqrt{\bar{\lambda}}_j t$$

$$= \frac{L}{2} - 2 \sum_{j=1}^{\infty} \frac{1}{\sqrt{\lambda}_j} \sin\sqrt{\bar{\lambda}}_j t. \qquad (1.1.49)$$

Let $u \epsilon L^2(0,T)$ be any solution of (1.1.44), (1.1.45), (1.1.23) for all $j \epsilon \mathbb{N}$. Then it follows from (1.1.49) that

$$\sum_{j=1}^{\infty} \frac{c_j^2}{\sqrt{\lambda}_j} = 0 \qquad (1.1.50)$$

Conversely let (1.1.50) hold. If we then put

$$u(t) = \frac{2}{L} \sum_{j=1}^{\infty} c_j^1 \cos\sqrt{\lambda_j}\, t + c_j^2 \sin\sqrt{\lambda_j}\, t, \tag{1.1.28}$$

we obtain a solution $u \in L^2[0,T]$ of (1.1.44), (1.1.45), (1.1.23) for all $j \in \mathbb{N}$. As a result we therefore have the

Theorem 1.1.6: Let $\mu = 0$ where μ is defined by (1.1.48) or, equivalently, let

$$\sum_{j=1}^{\infty} \frac{1}{\lambda_j} = \frac{L^2}{24}.$$

Then (1.1.44), (1.1.45), (1.1.23) for all $j \in \mathbb{N}$ and $T = L$ has a solution $u \in L^2(0,T)$, if and only if (1.1.50) is satisfied in which case u can be defined by (1.1.28) and has the smallest norm among all possible solutions of (1.1.44), (1.1.45), (1.1.23) for all $j \in \mathbb{N}$ in $L^2(0,T)$.

The last statement of the Theorem is again a consequence of Bessel's inequality.

Next we consider the case

b) $\mu > 0$, μ being defined by (1.1.48). Then the function

$$v(t) = -2 \sum_{j=1}^{\infty} \frac{1}{\sqrt{\lambda_j}} \sin\sqrt{\lambda_j}\, t + \frac{L}{2} \tag{1.1.51}$$

is in $L^2(0,L)$ because of (1.1.48) and is a solution of

$$\int_0^L v(t)\,dt = \frac{L^2}{2} = \int_0^L t\,dt$$

$$\int_0^L v(t)\cos\sqrt{\lambda_j}\, t\, dt = 0 = \int_0^L t\cos\sqrt{\lambda_j}\, t\, dt,$$

$$\int_0^L v(t)\sin\sqrt{\lambda_j}\, t\, dt = \int_0^L t\sin\sqrt{\lambda_j}\, t\, dt = -\frac{L}{\sqrt{\lambda_j}}$$

and

$$u(t) = \sum_{j=1}^{\infty} c_j^1 \cos\sqrt{\lambda_j}\, t + c_j^2 \sin\sqrt{\lambda_j}\, t$$
$$+ \left(\frac{L^2}{12} - \sum_{j=1}^{\infty} \frac{1}{\lambda_j}\right)^{-1} \left(\sum_{j=1}^{\infty} \frac{c_j^2}{\sqrt{\lambda_j}}\right)(t - v(t)) \tag{1.1.52}$$

is in $L^2(0,T)$ for $T = L$ and solves (1.1.44), (1.1.45) and (1.1.23) for all $j \in \mathbb{N}$.

Since u is in the closure of the span of $S \cup \{1\} \cup \{t\}$ in $L^2(0,L)$ with S defined by (1.1.24), it follows as in the proof of Theorem 1.1.5 that u is the only solution in this closure and has the smallest possible L^2-norm among all solutions of (1.1.44), (1.1.45) and (1.1.23) for all $j \in \mathbb{N}$ and $T = L$.

Next we consider the case $T > L$.

We observe that the solution $u \in L^2(0,T)$ of (1.1.23) for all $j \in \mathbb{N}$ also satisfies (1.1.45) as a consequence of the proof of Lemma 1.1.3 which implies that

$$\int_0^T u(t)dt = \int_0^L w(t)dt = 0$$

where w is defined by (1.1.33). Let again V be the closure of the span of $S \cup \{1\}$ in $L^2(0,T)$ with S defined by (1.1.24). Since all the functions in V are L-periodic, $z(t) = t$, $t \in [0,T]$, cannot belong to V. Therefore, by a well-known theorem in approximation theory, there is a unique $\hat{v} \in V$ with

$$0 < \| z - \hat{v} \|_{L^2(0,T)} \le \| z - v \|_{L^2(0,T)} \quad \text{for all } v \in V$$

which is characterized by

$$\int_0^T (z(t) - \hat{v}(t))v(t)dt = 0 \text{ for all } v \in V \qquad (1.1.53)$$

which is equivalent to

$$\int_0^T (z(t) - \hat{v}(t))dt = 0,$$

$$\int_0^T (z(t) - \hat{v}(t))\cos\sqrt{\lambda}_j t \, dt = 0,$$

$$\int_0^T (z(t) - \hat{v}(t))\sin\sqrt{\lambda}_j t \, dt = 0, \quad j \in \mathbb{N}.$$

Further it follows from (1.1.53) that

$$\alpha = \int_0^T (z(t)-\hat{v}(t))z(t)dt = \int_0^T (z(t)-\hat{v}(t))^2 dt > 0.$$

Let $u \in L^2(0,T)$ be any solution of (1.1.45) and (1.1.23) for all $j \in \mathbb{N}$ whose existence is guaranteed by the above considerations. Then

$$\hat{u}(t) = u(t) - \frac{1}{\alpha} \int_0^T s \cdot u(s)ds(t-\hat{v}(t)) \qquad (1.1.54)$$

is a solution of (1.1.44), (1.1.45) and (1.1.23) for all $j \in \mathbb{N}$. Since $\hat{u}$ is in the closure of the span of $S \cup \{1\} \cup \{t\}$ in $L^2(0,T)$, it again follows as in the proof of Theorem 1.1.5 that $\hat{u}$ is the only solution in this closure and has the smallest possible L^2-norm among all solutions of (1.1.44), (1.1.45) and (1.1.23) for all $j \in \mathbb{N}$.

Remark: For $T = L$ we have $z \notin V$, if and only if $\mu > 0$ with μ defined by (1.1.48), and the solution (1.1.52) of (1.1.44), (1.1.45), (1.1.23) for all $j \in \mathbb{N}$ and $T = L$ can be obtained as $\hat{u}$ (1.1.54). In fact v (1.1.54) is the unique solution $\hat{v}$ of (1.1.53).

Finally, we consider the case $T \in (0,L)$.

This can only be treated by elementary means, if we assume that S (1.1.24) $\cup \{1\}$ is complete. Then we are in the situation $\mu = 0$ with μ defined by (1.1.48), since the closure V of $S \cup \{1\}$ in $L^2(0,L)$ is the whole space $L^2(0,L)$ and $z(t) = t$, $t \in [0,L]$, necessarily belongs to V. As in the Corollary of Theorem 1.1.5 we now have the statement that $u \in L^2(0,T)$ being defined by (1.1.28) is the unique solution of (1.1.44), (1.1.45), (1.1.23) for all $j \in \mathbb{N}$, if (1.1.50) is satisfied and $u(t) = 0$ for almost all $t \in [T,L]$. Otherwise no solution exists.

We also conclude this subsection with the two special cases as in Section 1.1.2:

a) The Vibrating String.

In this case the differential equation (1.1.36) reads

$$y_{tt}(x,t) - y_{xx}(x,t) = 0 \qquad (1.1.36')$$

for $x \epsilon (0,1)$, $t > 0$, L is given by $Lz(x) = z''(x)$ with D_L defined by (1.1.8').

We consider boundary conditions of the form

$$B_1^0[y(\cdot,t)] = y(0,t) = 0,$$
$$B_1^1[y(\cdot,t)] = y(1,t) = v(t), \quad t \geq 0.$$

$$(1.1.37')$$

The solution $r \epsilon C^2[0,1]$ of (1.1.41) is given by

$$r(x) = x, \quad x \epsilon [0,1],$$

hence,

$$h_j = \int_0^1 r(x) e_j(x) dx = \sqrt{2} \int_0^1 x \sin j\pi x \, dx$$

$$= \sqrt{2} \frac{(-1)^{j+1}}{j\pi} \neq 0 \text{ for all } j \epsilon \mathbb{N},$$

i.e., (1.1.20) is satisfied.

Because of

$$\sum_{j=1}^{\infty} \lambda_j h_j^2 = 2 \sum_{j=1}^{\infty} (j\pi) = \infty$$

r does not belong to E (1.1.9).

Let us assume $y_0 \epsilon D_L$ (1.1.8') and $y_1 \epsilon H^1(0,1)$ with $y_1(0) = y_1(1) = 0$. Then

$$a_j = \int_0^1 y_0(x) e_j(x) dx = - \frac{\sqrt{2}}{(j\pi)^2} \int_0^1 y_0''(x) \sin j\pi x \, dx,$$

$$b_j = \int_0^1 y_1(x) e_j(x) dx = \frac{\sqrt{2}}{j\pi} \int_0^1 y_1'(x) \cos j\pi x \, dx, \quad j \epsilon \mathbb{N},$$

hence (1.1.27) is satisfied.

Moreover, we have

$$\sum_{j=1}^{\infty} \frac{1}{\lambda_j} = \sum_{j=1}^{\infty} \frac{1}{(j\pi)^2} = \frac{1}{6} = \frac{L^2}{24},$$

i.e., we are in the situation $\mu = 0$ with μ being defined by (1.1.48) which also follows from the fact that

$$S_u\{1\} = \{1, \cos j\pi t, \sin j\pi t \mid t \geq 0, j \epsilon \mathbb{N}\}$$

is complete on $[0,2]$.

So the above results for $\mu = 0$ can be applied where the condition (1.1.50) reads

$$\sum_{j=1}^{\infty} \frac{(-1)^{j+1}}{j\pi} \int_0^1 y_0''(x)\sin j\pi x \, dx = 0. \qquad (1.1.50')$$

It can be expressed more directly as follows: If we extend y_0 as an odd 2-periodic function on $\mathbb{R}$, then y_0' has a Fourier series expansion of the form

$$y_0'(x) = \frac{1}{2} \int_0^2 y_0'(\xi)d\xi + \sum_{j=1}^{\infty} \int_0^2 y_0'(\xi)\cos j\pi\xi \, d\xi \, \cos j\pi x$$

$$+ \int_0^2 y_0'(\xi)\sin j\pi\xi d\xi \, \sin j\pi x$$

$$= \sum_{j=1}^{\infty} - \frac{1}{j\pi} \int_0^2 y_0''(\xi)\sin j\pi\xi d\xi\cos j\pi x, \quad x\in[0,1].$$

Therefore (1.1.50') turns out to be equivalent to $y_0'(1) = 0$ and for $T = 2$ a solution of (1.1.44), (1.1.45), (1.1.23) for all $j\in \mathbb{N}$ exists, if and only if $y_0'(1) = 0$ in which case

$$u(t) = \sum_{j=1}^{\infty} c_j^1\cos j\pi t + c_j^2\sin j\pi t, \quad t\in[0,2] \qquad (1.1.28')$$

is a least norm solution in $L^2(0,2)$.

If $T>2$, there exists a least norm solution of (1.1.44), (1.1.45), (1.1.23) for all $j\in \mathbb{N}$. If $T\in(0,2)$, u is given by (1.1.28') is the only solution of (1.1.44), (1.1.45), (1.1.23) for all $j\in \mathbb{N}$, if and only if $u(t) = 0$ for all $t\in(T,2]$ and $y_0'(1) = 0$.

b) The Vibrating (Euler-) Beam.

In this case the differential equation (1.1.36) reads

$$y_{tt}(x,t) + y_{xxxx}(x,t) = 0 \qquad (1.1.36'')$$

for $x\in(0,1)$ and $t>0$. We consider boundary conditions (1.1.37') of the form

$$B_1^0[y(\cdot,t)] = y(0,t)= 0, \quad B_1^1[y(\cdot,t)] = y(1,t) = v(t),$$
$$B_2^0[y(\cdot,t)] = y_{xx}(0,t) = 0, \quad B_2^1[y(\cdot,t)] = y_{xx}(1,t) = 0, \quad t\geq 0. \qquad (1.1.37'')$$

So $Lz = z^{(4)}$ and D_L is defined by (1.1.8").

The solution $r \epsilon C^4[0,1]$ of (1.1.41) is also given by

$$r(x) = x, \quad x \epsilon [0,1],$$

hence again

$$h_j = \sqrt{2} \frac{(-1)^{j+1}}{j\pi} \neq 0 \text{ for all } j \epsilon \mathbb{N}$$

and (1.1.20) is satisfied.

We assume $y_0 \epsilon D_L$ (1.1.8") and $y_1 \epsilon H^1(0,1)$ with $y_1(0) = y_1(1) = 0$.
Then

$$a_j = \int_0^1 y_0(x) e_j(x) dx = \frac{\sqrt{2}}{(j\pi)^4} \int_0^1 y_0^{(4)}(x) \sin j\pi x \, dx,$$

$$b_j = \int_0^1 y_1(x) e_j(x) dx = \frac{\sqrt{2}}{j\pi} \int_0^1 y_1'(x) \sin j\pi x dx, \quad j \epsilon \mathbb{N},$$

hence (1.1.27) is satisfied.

For $L = \frac{2}{\pi}$ the system

$$S = \{\cos\sqrt{\lambda_j} t, \sin\sqrt{\lambda_j} t \mid t \geq 0, \ j \epsilon \mathbb{N}\}$$

$$= \{\cos(j\pi)^2 t, \sin(j\pi)^2 t \mid t \geq 0, \ j \epsilon \mathbb{N}\}$$

is orthogonal on $[0,L]$ and L-periodic. However, $S \cup \{1\}$ is not complete in $L^2(0,L)$ which is also a consequence of

$$\mu = \frac{L^3}{12} - 2L \sum_{j=1}^{\infty} \frac{1}{\lambda_j} = \frac{2}{3\pi^3} - \frac{4}{\pi} \sum_{j=1}^{\infty} \frac{1}{(j\pi)^4}$$

$$= \frac{2}{3\pi^3} - \frac{4}{\pi^5} \cdot \frac{\pi^4}{90} = \frac{2}{3\pi} (\frac{1}{\pi^2} - \frac{1}{15}) > 0.$$

So, by the above results, for every $T \geq \frac{2}{\pi}$, there is a solution $u \epsilon L^2(0,T)$ of (1.1.44), (1.1.45) and (1.1.23) for all $j \epsilon \mathbb{N}$.

For $T \epsilon (0, \frac{2}{\pi})$ no statement can be made so far. But we shall see in Section 1.2.3.2 that the existence of a solution of (1.1.44), (1.1.45) and (1.1.23) for all $j \epsilon \mathbb{N}$ can also be proved in this case.

1.2 On Moment Problems in Hilbert Spaces.

1.2.1 Problems in General Hilbert Spaces.

Let Z be a Hilbert space over the real or complex numbers whose scalar product is denoted by $\langle \cdot, \cdot \rangle$. A moment problem in Z is defined by considering a sequence $(z_j)_{j \in \mathbb{N}}$ in Z and a sequence $(c_j)_{j \in \mathbb{N}}$ of real or complex numbers and by asking for some $u \in Z$ such that

$$\langle u, z_j \rangle = c_j, \quad j \in \mathbb{N}, \tag{1.2.1}$$

is satisfied.

We assume that the sequence $(z_j)_{j \in \mathbb{N}}$ is linearly independent in Z, i.e., every finite subsequence $(z_j)_{j=1,\ldots,N}$, $N \in \mathbb{N}$, of the sequence $(z_j)_{j \in \mathbb{N}}$ is linearly independent. Then we can prove

Theorem 1.2.1: For every $N \in \mathbb{N}$ there is exactly one solution $u = u^N \in Z$ of

$$\langle u, z_j \rangle = c_j, \quad j \in \{1,\ldots,N\}, \tag{1.2.1$_N$}$$

of the form

$$u^N = \sum_{j=1}^{N} \xi_j^N z_j \tag{1.2.2}$$

and u^N has the smallest possible norm $|u^N|$ among all solutions of $(1.2.1)_N$ in Z.

Proof: An element $u^N \in Z$ of the form (1.2.2) is a solution of $(1.2.1)_N$, if and only if

$$\sum_{i=1}^{N} \langle z_i, z_j \rangle \xi_i^N = c_j, \quad j \in \{1,\ldots,N\}. \tag{1.2.3}$$

Due to the linear independence of $z_1, \ldots, z_n$, Gram's matrix

$$G_N = (\langle z_i, z_j \rangle)_{i,j=1,\ldots,N} \tag{1.2.4}$$

is Hermitian and positive definite so that (1.2.3) has a unique solution $(\xi_1^N, \ldots, \xi_N^N) \in \mathbb{R}^N$ or $\mathbb{C}^N$. Therefore there is exactly one solution $u = u^N$ of $(1.2.1)_N$ which is of the form (1.2.2) with

$$\xi_j^N = \sum_{k=1}^{N} \sigma_{jk}^N c_k, \quad j \in \{1,\ldots,N\}, \tag{1.2.5}$$

where

$$\sigma_{j,k}^N = (G_N^{-1})_{j,k}, \quad j,k \in \{1,\ldots,N\} \tag{1.2.6}$$

Let $u \in Z$ be any solution of (1.2.1). Then

$$<u - u^N, z_j> = 0 \text{ for all } j = 1, \ldots, N$$

and hence

$$<u - u^N, u^N> = 0$$

which implies

$$0 \leq <u - u^N, u - u^N> = |u|^2 - |u^N|^2$$

and completes the proof.

Remark: Due to the strict convexity of the functional $u \to |u|^2$, $u \in Z$, the element u^N (1.2.2) with (1.2.5), (1.2.6) is also the unique least norm solution of $(1.2.1)_N$.

For each $N \in \mathbb{N}$ we define

$$\lambda_N(c) = \inf\{|u| \mid u \in Z, u \text{ solves } (1.2.1)_N\}. \tag{1.2.7}$$

Apparently we have

$$\lambda_N(c) \leq \lambda_{N+1}(c) \text{ for all } N \in \mathbb{N}. \tag{1.2.8}$$

If $u \in Z$ is a solution (1.2.1), then u solves $(1.2.1)_N$ for every $N \in \mathbb{N}$ and therefore, by Theorem 1.2.1,

$$\lambda_N(c) \leq |u| \text{ for all } N \in \mathbb{N} \tag{1.2.9}$$

Conversely we have the following

Theorem 1.2.2: Let Z be a separable Hilbert space.

If

$$\lambda = \lim_{N \to \infty} \lambda_N(c) < \infty, \tag{1.2.10}$$

then there is exactly one solution $u = u^\infty \in Z$ of (1.2.1) with least norm which is given by

$$u^\infty = \lim_{N \to \infty} u^N \tag{1.2.11}$$

where each $u = u^N$ is the unique least norm solution of $(1.2.1)_N$ (which moreover is of the form (1.2.2), (1.2.5), (1.2.6)).

Proof: Since Z is separable, the set $B_\lambda = \{u \in Z \mid |u| \le \lambda\}$
is weakly sequentially compact. Therefore the sequence
$(u^N)_{N \in \mathbb{N}}$ of unique least norm solutions $u = u^N$ of $(1.2.1)_N$
as guaranteed by Theorem 1.2.1 has a subsequence $(u^{N_i})_{i \in \mathbb{N}}$
which weakly converges to some $u^\infty \in B_\lambda$. This implies

$$\langle u^\infty, z_j \rangle = c_j \text{ for all } j \in \mathbb{N}$$

and hence $\lambda \le |u^\infty|$ because of (1.2.9), (1.2.10). Hence $|u^\infty| = \lambda$
because of $u \in B_\lambda$ which shows that u^∞ is a least norm solution
of (1.2.1). Again from the strict convexity of the functional
$u \to |u|^2$, $u \in Z$, it follows that u^∞ is the unique least norm
solution of (1.2.1). This in turn implies that the whole
sequence $(u^N)_{N \in \mathbb{N}}$ converges weakly to u^∞. Since $|u^N| = \lambda_N(c)$
for all $N \in \mathbb{N}$, we also have by (1.2.10) that

$$|u^\infty| = \lambda = \lim_{N \to \infty} |u^N|$$

from which finally (1.2.11) follows.

Corollary: Under the assumptions of Theorem 1.2.2 the least norm
solution $u^\infty \in Z$ of (1.2.1) which is given by (1.2.11) is the unique
solution of (1.2.1) in the closure of the span of $S = \{z_j \mid j \in \mathbb{N}\}$.

Proof: From (1.2.11) it follows that $u^\infty \in V = $ closure of the span
of S. Let $u \in V$ be arbitrary and a solution of (1.2.1). Then

$$\langle u^\infty - u, z_j \rangle = 0 \text{ for all } j \in \mathbb{N}$$

which implies

$$|u^\infty - u|^2 = \langle u^\infty - u, u^\infty - u \rangle = 0 \Longrightarrow u^\infty = u.$$

A simple consequence of these results is in the

Theorem 1.2.3: Let Z be a separable Hilbert space. If there
exists a constant $\mu > 0$ such that

$$\sum_{j=1}^{N} |\xi_j|^2 \le \mu |\sum_{j=1}^{N} \xi_j z_j|^2 \tag{1.2.12}$$

for all $N \in \mathbb{N}$ and all $(\xi_1, \ldots, \xi_N) \in \mathbb{R}^N$ or $\mathbb{C}^N$, then for all sequences
$c = (c_j)_{j \in \mathbb{N}}$ with

$$\sum_{j=1}^{N} |c_j|^2 < \infty, \qquad\qquad (1.2.13)$$

the condition (1.2.10) is satisfied and, consequently, (1.2.1) has a unique least norm solution $u = u^\infty$ which is given by (1.2.11).

For the proof we need

Lemma 1.2.4: For the solution u^N (1.2.2), (1.2.5), (1.2.6) of $(1.2.1)_N$ we have

$$\|u^N\| = \sup\{ | \sum_{j=1}^{N} c_j \overline{\xi_j}| \mid | \sum_{j=1}^{N} \xi_j z_j| \le 1\}. \qquad (1.2.14)$$

Proof: Let $u \in Z$ be any solution of $(1.2.1)_N$ and let $(\xi_1,\ldots,\xi_N) \in \mathbb{R}^N$ or $\mathbb{C}^N$ be given such that

$$| \sum_{j=1}^{N} \xi_j z_j| \le 1.$$

Then

$$| \sum_{j=1}^{N} c_j \overline{\xi_j}| = | \sum_{j=1}^{N} <u,z_j> \overline{\xi_j}| = |<u, \sum_{j=1}^{N} \xi_j z_j>| \le \|u\|$$

and, consequently,

$$\sup\{ | \sum_{j=1}^{N} c_j \overline{\xi_j}| \mid | \sum_{j=1}^{N} \xi_j z_j| \le 1\} \le \lambda_N(c) \ (1.2.7), \qquad (1.2.15)$$

since u and $(\xi_1,\ldots,\xi_N)$ are chosen arbitrarily.

For u^N defined by (1.2.2), (1.2.5), (1.2.6) it follows that

$$| \sum_{j=1}^{N} c_j \overline{\xi_j^N}| = \sum_{j=1}^{N} <u^N,z_j> \overline{\xi^N} = <u^N,u^N> = \|u^N\|^2$$

and therefore

$$| \sum_{j=1}^{N} c_j \overline{\xi_j^*}| = \|u^N\|, \quad | \sum_{j=1}^{N} \xi_j^* z_j| = 1, \qquad (1.2.16)$$

if we define

$$\xi_j^* = \frac{\xi_j^N}{\|u^N\|} \quad \text{for } j = 1,\ldots,N$$

in the case where $|u^N|>0$ (in the case $|u^N| = 0$ the assertion
(1.2.14) follows trivially from (1.2.15)). Hence (1.2.14) follows
from (1.2.15), (1.2.16).

Proof of Theorem 1.2.3: For each $N \in \mathbb{N}$ and $(\xi_1, \ldots, \xi_N) \in \mathbb{R}^N$ or $\mathbb{C}^N$
it follows with (1.2.12) that

$$| \sum_{j=1}^{N} c_j \overline{\xi_j} | \leq (\sum_{j=1}^{N} |c_j|^2)^{1/2} (\sum_{j=1}^{N} |\xi_j|^2)^{1/2} \leq \beta | \sum_{j=1}^{N} \xi_j z_j |$$

where

$$\beta = \mu^{1/2} (\sum_{j=1}^{\infty} |c_j|^2)^{1/2} < \infty \qquad (1.2.17)$$

by (1.2.13). As a consequence of Lemma 1.2.4 we therefore have

$$\lambda_N(c) = |u^N| \leq \beta$$

which implies (1.2.10) and concludes the proof.

Corollary: Under the assumptions of Theorem 1.2.3 the norm of
the unique least norm solution $u = u^\infty$ (1.2.11) of (1.2.1) for
a given sequence $(c_j)_{j \in \mathbb{N}}$ with (1.2.13) satisfies the inequality

$$|u^\infty| \leq \beta \qquad (1.2.18)$$

with β defined by (1.2.17).

Under an additional assumption on the sequence $(z_j)_{j \in \mathbb{N}}$ the
statement of Theorem 1.2.3 can be strengthened. For this purpose
we first give the

Definition: A sequence $(z_j)_{j \in \mathbb{N}}$ in Z is called minimal, if, for
each $k \in \mathbb{N}$, the element z_k does not belong to closure of the span
of $S_k = \{z_j | j \in \mathbb{N}, j \neq k\}$.

As an immediate consequence of the definition we see that every
minimal sequence in Z is linearly independent. The converse is
false in general. Minimality can be considered as a strong form
of linear independence.

Useful is the following characterization of minimality which we
formulate as

Theorem 1.2.5: A sequence $(x_j)_{j \in \mathbb{N}}$ in Z is minimal, if and only if there exists a so called biorthonormal sequence, i.e., a sequence $(y_k)_{k \in \mathbb{N}}$ in Z with

$$\langle y_k, x_j \rangle = \delta_{kj} \text{ for all } j, k \in \mathbb{N}, \tag{1.2.19}$$

δ_{kj} being Kronecker's symbol.

Proof: 1) Let $(x_j)_{j \in \mathbb{N}}$ be minimal in Z. For every $k \in \mathbb{N}$ we put V_k = closure of the span of S_k. Then from $x_k \notin V_k$ it follows from a well-known theorem in approximation theory that there is a unique $v_k \in V_k$ with

$$0 < |x_k - v_k| \le |x_k - v| \text{ for all } v \in V_k$$

which is characterized by

$$\langle x_k - v_k, v \rangle = 0 \text{ for all } v \in V_k,$$

in particular

$$\langle x_k - v_k, x_j \rangle = 0 \text{ for all } j \neq k \text{ in } N.$$

If we put $y_k = \dfrac{x_k - v_k}{\|x_k - v_k\|^2}$, then (1.2.19) follows.

2) Let (1.2.19) be satisfied. If, for some $k \in \mathbb{N}$, we assume $x_k \in V_k$, then (1.2.19) implies

$$\langle y_k, x_k \rangle = 1 \text{ and } \langle y_k, x_k \rangle = 0$$

which is a contradiction.

Remark: By the proof of Theorem 1.2.5 one can assume the sequence $(y_k)_{k \in \mathbb{N}}$ with (1.2.19) to belong to the closure of the span of $\{x_j \mid j \in N\}$.

The strengthened form of Theorem 1.2.3 now reads as follows.

Theorem 1.2.6: Let Z be a separable Hilbert space and let $(z_j)_{j \in \mathbb{N}}$ be minimal in Z. If there exists a constant $\mu > 0$ such that $\displaystyle\sum_{j=N_0}^{N} |\xi_j|^2 \le \mu \left\| \sum_{j=N_0}^{N} \xi_j z_j \right\|^2$ or $\mathbb{C}^{N-N_0+1}$ for all $N \ge N_0$ and all $(\xi_{N_0}, \ldots, \xi_N) \in \mathbb{R}^{N-N_0+1}$ where $N_0 \in \mathbb{N}$ is fixed then for all sequences $(c_j)_{j \in \mathbb{N}}$ with (1.2.13) there exists a unique least norm solution of (1.2.1) in the closure of the span of $S = \{z_j \mid j \in \mathbb{N}\}$.

Proof: Let $(c_j)_{j \in \mathbb{N}}$ with (1.2.13) be given. Then, by Theorem 1.2.3 there exists a solution $u \in Z$ of

$$\langle u, z_j \rangle = c_j \text{ for } j \geq N_0$$

which is in the closure of the span of $\{z_j | j \geq N_0\}$. Let $(y_k)_{k \in \mathbb{N}}$ be any sequence in the closure of the span of S with

$$\langle y_k, z_j \rangle = \delta_{kj} \quad \text{for all } j, k \in N \tag{1.2.19'}$$

whose existence is ensured by the remark following Theorem 1.2.5. If we put

$$u^{\infty} = u + \sum_{j=1}^{N_0-1} (c_j - \langle u, z_j \rangle) y_j, \tag{1.2.20}$$

then, by the proof of the Corollary of Theorem 1.2.2, u^{∞} is the unique solution of (1.2.1) in the closure of the span of S and therefore also the unique least norm solution of (1.2.1).

If $(z_j)_{j \in \mathbb{N}}$ is minimal, a formal solution of (1.2.1) can be written down immediately as

$$u^{\infty} = \sum_{j=1}^{\infty} c_j y_j \tag{1.2.21}$$

where $(y_j)_{j \in \mathbb{N}}$ is any sequence with (1.2.19') whose existence is ensured by Theorem 1.2.5. The question, however, is whether the series in (1.2.21) converges in Z so that u^{∞} is in Z or even in the closure of the span of $S = \{z_j | j \in \mathbb{N}\}$.

If $(z_j)_{j \in \mathbb{N}}$ is orthonormal, then the sequence $(y_j)_{j \in \mathbb{N}}$ can be chosen to be $(z_j)_{j \in \mathbb{N}}$ itself and for every sequence $(c_j)_{j \in \mathbb{N}}$ with (1.2.13) it follows that

$$u^{\infty} = \sum_{j=1}^{\infty} c_j z_j$$

is the unique solution of (1.2.1) in the closure of the span of S.

This result can be generalized. For that purpose we first prove the

Lemma 1.2.7: Let $(z_j)_{j \in \mathbb{N}}$ be a sequence in Z such that the condition (1.2.12) is satisfied for all $N \in \mathbb{N}$ and all $(\xi_1, \ldots, \xi_N) \in \mathbb{R}^N$ or $\mathbb{C}^N$. Then $(z_j)_{j \in \mathbb{N}}$ is minimal.

Proof: We assume that, for some $k \in \mathbb{N}$, the element z_k is in the closure of the span of $\{z_j \mid j \in \mathbb{N}, j \neq k\}$. Then

$$z_k = \lim_{N \to \infty} \sum_{\substack{j=1 \\ j \neq k}}^{N} \xi_j^N z_j.$$

From (1.2.12) it follows that

$$\mu \left| z_k - \sum_{\substack{j=1 \\ j \neq k}}^{N} \xi_j^N z_j \right|^2 \geq 1 + \sum_{\substack{j=1 \\ j \neq k}}^{N} |\xi_j^N|^2 \geq 1$$

which contradicts

$$\lim_{N \to \infty} \left| z_k - \sum_{\substack{j=1 \\ j \neq k}}^{N} \xi_j^N z_j \right| = 0.$$

This completes the proof of the minimality of the sequence $(z_j)_{j \in \mathbb{N}}$.

With this result we can prove the

Theorem 1.2.8: Under the assumptions of Theorem 1.2.3 it follows for every sequence $(c_j)_{j \in \mathbb{N}}$ with (1.2.13) that the unique least norm solution of (1.2.1) can be represented by (1.2.21) where $(y_j)_{j \in \mathbb{N}}$ is any sequence in the closure of the span of $S = \{z_j \mid j \in \mathbb{N}\}$ with (1.2.19') (guaranteed by Theorem 1.2.5 and Lemma 1.2.7).

Proof: For every $N \in \mathbb{N}$ we define $u_N = \sum_{j=1}^{N} c_j y_j$. Then

$$\langle u_N, z_j \rangle = c_j \quad \text{for } j = 1, \ldots, N$$

and

$$\langle u_N, z_j \rangle = 0 \quad \text{for all } j \geq N + 1.$$

If u_N^∞ is the unique least norm solution of

$$\langle u_N^\infty, z_j \rangle = 0 \quad \text{for } j = 1, \ldots, N,$$

$$\langle u_N^\infty, z_j \rangle = c_j \quad \text{for all } j \geq N + 1$$

whose existence is ensured by Theorem 1.2.3. By the Corollary of Theorem 1.2.3 we further have that

$$|u_N^\infty| \le \mu^{1/2} (\sum_{j=N+1}^{\infty} |c_j|^2)^{1/2} .$$

Since $u = u_N + u_N^\infty$ is a solution of (1.2.1) which is in the closure of the span of S it is the unique least norm solution u^∞ of (1.2.1), by the Corollary of Theorem 1.2.2. Hence

$$|u^\infty - u_N| = |u_N^\infty| \le \mu^{1/2} (\sum_{j=N+1}^{\infty} |c_j|^2)^{1/2}$$

and therefore

$$u^\infty = \lim_{N \to \infty} u_N = \sum_{j=1}^{\infty} c_j y_j$$

by virtue of (1.2.13). This concludes the proof.

Under the additional assumption that there exists a constant $\delta > 0$ such that

$$| \sum_{j=1}^{N} \xi_j z_j |^2 \le \delta \sum_{j=1}^{N} |\xi_j|^2 \tag{1.2.22}$$

for all $N \in \mathbb{N}$ and all $(\xi_1, \ldots, \xi_N) \in \mathbb{R}^N$ or $\mathbb{C}^N$ one can also prove that the unique least norm solution $u^\infty \in Z$ of (1.2.1) can be represented in the form

$$u^\infty = \sum_{j=1}^{\infty} a_j z_j$$

for some sequence $(a_j)_{j \in \mathbb{N}} \in l_2$.

1.2.2. Trigonometric Moment Problems.

In order to deal with the moment problem (1.1.23) for the case of distributed control where the system S (1.1.24) is not L-periodic for some $L > 0$ and orthogonal on $[0, L]$ we consider at first the following complex trigonometric moment problem which consists, for some given $T > 0$ and a sequence $(c_j)_{j \in \mathbb{N}}$ with (1.2.13), of finding a function $u \in L^2(0, T)$ such that

$$\int_0^T u(t)e^{2i\omega_j t} \, dt = c_{2j-1},$$

$$\int_0^T u(t)e^{-2i\omega_j t} \, dt = c_{2j}, \quad j \in \mathbb{N}\, (i = \sqrt{-1}). \tag{1.2.24}$$

We further assume that $(\omega_j)_{j \in \mathbb{N}}$ is a strictly increasing sequence of positive reals.

If we put

$$z_{2j-1}(t) = e^{-2i\omega_j t} \text{ and } z_{2j}(t) = e^{2i\omega_j t}, \quad t \geq 0, \tag{1.2.25}$$

then we have a moment problem of the form (1.2.1) in $Z = L^2(0,T)$ equipped with the scalar product

$$\langle u, v \rangle = \int_0^T u(t)\, \overline{v(t)}\, dt, \quad u, v \in L^2(0,T)$$

where the bar denotes the conjugate complex.

By the assumption on the sequence $(\omega_j)_{j \in \mathbb{N}}$ the sequence $(z_j)_{j \in \mathbb{N}}$ is linearly independent. In order to be able to apply Theorem 1.2.3 to the moment problem (1.2.24) we have to make sure that the condition (1.2.12) holds. For this purpose we prove

<u>Theorem 1.2.9</u>: Let $\underline{a}_N, \dots, \underline{a}_1, a_0, a_1, \dots, a_N$ arbitrarily chosen complex numbers and let $\underline{\omega}_N, \dots, \underline{\omega}_1, \omega_0, \omega_1, \dots, \omega_N$ be reals such that

$$\omega_j - \omega_{j-1} \geq \lambda \quad \text{for } -N < j \leq N \tag{1.2.26}$$

and some $\lambda > 0$. Then for each $\epsilon > 0$ and for $T = \frac{\pi + \epsilon}{\lambda}$ it follows that

$$\sum_{j=-N}^{N} |a_j|^2 \leq \frac{A(\epsilon)}{2T} \int_{-T}^{T} |\sum_{j=-N}^{N} a_j \, e^{-i\omega_j t}|^2 \, dt \tag{1.2.27}$$

where

$$A(\epsilon) = \frac{\pi (\pi+\epsilon)^2}{2\epsilon (2\pi+\epsilon)}. \tag{1.2.28}$$

<u>Proof:</u> Let

$$k(t) = \begin{cases} \cos \dfrac{t}{2} & \text{for } |t| \leq \pi, \\ 0 & \text{for } |t| > \pi. \end{cases}$$

If we put

$$K(\omega) = \int_{-\infty}^{\infty} k(t)\, e^{-i\omega t}\, dt,$$

then for

$$f(t) = \sum_{j=-N}^{N} a_j\, e^{-i\omega_j t}$$

we obtain

$$\int_{-\infty}^{\infty} k(t)\,|f(t)|^2\, dt = \sum_{j,k=-N}^{N} a_j\, \overline{a_k}\, K(\omega_j - \omega_k)$$

where

$$K(\omega) = \frac{4\cos\omega\pi}{1 - 4\omega^2} = K(-\omega) \quad \text{for all } \omega \neq \pm\frac{1}{2}$$

and

$$K(\pm\tfrac{1}{2}) = \pi.$$

This implies

$$\int_{-\infty}^{\infty} k(t)\,|f(t)|^2\, dt = \sum_{j=-N}^{N} |a_j|^2 K(0) + \sum_{j=-N}^{N} \sum_{\substack{k=-N \\ k\neq j}}^{N} (a_j\overline{a_k})\, K(\omega_j - \omega_k)$$

$$\geq 4 \sum_{j=-N}^{N} |a_j|^2 - \sum_{j=-N}^{N} \sum_{\substack{k=-N \\ k\neq j}}^{N} \frac{|a_j|^2 + |a_k|^2}{2}\, |\,K(\omega_j - \omega_k)\,|$$

$$= \sum_{j=-N}^{N} |a_j|^2 \left(4 - \frac{1}{2}\sum_{\substack{k=-N \\ k\neq j}}^{N} |K(\omega_j - \omega_k)|\right) - \sum_{j=-N}^{N} \sum_{\substack{k=-N \\ k\neq j}}^{N} \frac{|a_k|^2}{2}\, |\,K(\omega_j - \omega_k)\,|$$

$$\geq \sum_{j=-N}^{N} |a_j|^2 \left(4 - \frac{1}{2}\sum_{\substack{k=-N \\ k=j}}^{N} |K(\omega_j - \omega_k)|\right) - \sum_{j=-N}^{N} \frac{1}{2}|a_j|^2 \sum_{\substack{k=-N \\ k\neq j}}^{N} |K(\omega_k - \omega_j)|$$

$$\geq 2 \sum_{j=-N}^{N} |a_j|^2 \left(2 - \sum_{\substack{k=-N \\ k\neq j}}^{N} |K(\omega_j - \omega_k)|\right)$$

Because of

$$\frac{1}{2T} \int_{-T}^{T} |f(t)|^2 \, dt = \frac{2}{2\pi} \int_{-\pi}^{\pi} |f(\frac{T}{\pi}s)|^2 \, ds$$

$$= \frac{1}{2\pi} \int_{-\pi}^{\pi} | \sum_{j=-N}^{N} a_j \, e^{-i\frac{\omega_j T}{\pi}s} |^2 \, ds$$

and

$$\frac{\omega_j T}{\pi} - \frac{\omega_{j-1} T}{\pi} > \lambda \frac{T}{\pi} = \frac{\pi + \epsilon}{\pi} > 1$$

we can assume $T = \pi$ and $\lambda = \frac{\pi + \epsilon}{\pi}$.

For each pair $j, k \in \{-N, \ldots, N\}$ with $j \neq k$ it then follows that

$$|\omega_j - \omega_k| > |j-k| \, \lambda > 1$$

and therefore

$$|K(\omega_j - \omega_k)| \leq \frac{4}{4(j-k)^2\lambda^2 - 1} < \frac{4}{\lambda^2} \frac{1}{4(j-k)^2 - 1} \, .$$

For each $j \in \{-N, \ldots, N\}$ we therefore conclude that

$$\sum_{\substack{k=-N \\ k \neq j}}^{N} |K(\omega_j - \omega_k)| < \frac{2}{\lambda^2} \sum_{n=1}^{\infty} \frac{2}{4n^2 - 1} = \frac{2}{\lambda^2} \sum_{n=1}^{\infty} (\frac{1}{2n-1} - \frac{1}{2n+1}) = \frac{2}{\lambda^2}$$

As a result we obtain

$$\int_{-\pi}^{\pi} |f(t)|^2 \, dt \geq \int_{-\pi}^{\pi} k(t) |f(t)|^2 \, dt$$

$$\geq 2 \sum_{j=-N}^{N} |a_j|^2 (2 - \frac{2}{\lambda^2}) = 2 \sum_{j=-N}^{N} |a_j|^2 \frac{2\epsilon(2\pi + \epsilon)}{(\pi + \epsilon)^2}$$

or

$$\sum_{j=-N}^{N} |a_j|^2 \leq \frac{\pi(\pi + \epsilon)^2}{2\epsilon(2\pi + \epsilon)} \frac{1}{2\pi} \int_{-\pi}^{\pi} |f(t)|^2 \, dt$$

which is exactly the assertion (1.2.27), (1.2.28) for $T = \pi$ and therefore completes the proof.

As an easy consequence we obtain the

Lemma 1.2.10: Let $\omega_0 = 0$ and assume that there exists a real number $\epsilon > 0$ such that

$$\omega_j - \omega_{j-1} \geq \frac{\pi + \epsilon}{T} \quad \text{for all } j \in \mathbb{N}. \tag{1.2.29}$$

Then there exists a constant $m(T,\epsilon) > 0$ such that

$$m(T,\epsilon) \sum_{j=0}^{N} |a_j|^2 \leq \int_0^T |\sum_{j=0}^{N} a_j z_j(t)|^2 \, dt \tag{1.2.30}$$

for all $N \in \mathbb{N}$ and all $(a_0, \ldots, a_N) \in \mathbb{R}^{N+1}$ or $\mathbb{C}^{N+1}$ where $(z_j)_{j \in \mathbb{N}}$ is defined by (1.2.25) and $z_0 \equiv 1$.

Proof: It suffices to prove (1.2.30) for all $N = 2n$, $n \in \mathbb{N}$. If we define $\underline{\omega}_j = -\omega_j$ for all $j \in \mathbb{N}$, then (1.2.29) is true for all $j \in \mathbb{Z}$. By Theorem 1.2.9 it follows, for every $n \in \mathbb{N}$ and every choice of $(\underline{b}_n, \ldots, b_n) \in \mathbb{R}^{2n+1}$ or $\mathbb{C}^{2n+1}$ that

$$\sum_{j=-n}^{n} |b_j|^2 \leq \frac{A(\epsilon)}{2T} \int_{-T}^{T} |\sum_{j=-n}^{n} b_j e^{-i\omega_j t}|^2 \, dt$$

where $A(\epsilon)$ is given by (1.2.28).

Let $N = 2n$, $n \in \mathbb{N}$, and $(a_0, \ldots, a_N) \in \mathbb{R}^{N+1}$ or $\mathbb{C}^{N+1}$ be given. Then we define

$$b_0 = a_0, \quad b_j = a_{2j-1} e^{i\omega_j T}, \quad \underline{b}_j = a_{2j} e^{-i\omega_j T}$$

and conclude by virtue of Theorem 1.2.9 that

$$\sum_{j=0}^{N} |a_j|^2 = \sum_{j=-1}^{n} |b_j|^2 \leq \frac{A(\epsilon)}{2T} \int_{-T}^{T} |\sum_{j=-n}^{n} b_j e^{-i\omega_j t}|^2 \, dt$$

$$= \frac{A(\epsilon)}{T} \int_0^T |\sum_{j=-n}^{n} b_j e^{i\omega_j T} e^{-2i\omega_j t}|^2 \, dt$$

$$= \frac{A(\epsilon)}{T} \int_0^T |\sum_{j=0}^{N} a_j z_j(t)|^2 \, dt$$

which completes the proof.

By virtue of Lemma 1.2.7 we get the

Corollary: Under the assumption of Lemma 1.2.10 the sequence
$\{z_j\}_{j\in \mathbb{N}\cup\{0\}}$ with z_j defined by (1.2.25) for $j\epsilon$ $\mathbb{N}$ and $z_0 \equiv 1$
is minimal on [0,T].

On using Theorems 1.2.3 and 1.2.8 we obtain as first main
result the

Theorem 1.2.11: Under the assumption of Lemma 1.2.10 there is,
for every sequence $(c_j)_{j\epsilon \mathbb{N}}$ with (1.2.13) exactly one minimum
norm solution $u^\infty \epsilon L^2(0,T)$ of (1.2.24) which is also the unique
solution of (1.2.24) in the closure of the span of
$S = \{e^{-2i\omega_j t}, e^{2i\omega_j t} \mid t\epsilon[0,T], j\epsilon \mathbb{N}\}$ and has a representation

$$u^\infty(t) = \sum_{j=1}^{\infty} c_j y_j(t)$$

where $(y_j)_{j\epsilon \mathbb{N}}$ is any sequence in the closure of the span of S
with

$$\int_0^T y_k(t) z_j(t) \, dt = \delta_{kj} \quad \text{for all } j, k\epsilon \mathbb{N}$$

and $(z_j)_{j\epsilon \mathbb{N}}$ being defined by (1.2.25).

The Corollary of Theorem 1.2.3 leads to the following

Corollary: Under the assumption of Lemma 1.2.10 the norm of the
unique minimum norm solution $u^\infty \epsilon L^2(0,T)$ of (1.2.24) can be
estimated by

$$|u^\infty| \le \left(\frac{A(\epsilon)}{T}\right)^{1/2} \left(\sum_{j=1}^{\infty} |c_j|^2\right)^{1/2}$$

From the Theorem 1.2.3 and 1.2.8 we can also deduce the

Theorem 1.2.12: Under the assumptions of Lemma 1.2.10 there is,
for every sequence $(c_j)_{j\epsilon \mathbb{N}}$ with (1.2.13), exactly one minimum
norm solution $u \epsilon L^2(0,T)$ of (1.2.24) and

$$\int_0^T u(t) \, dt = c_0 \tag{1.2.31}$$

for any $c_0\epsilon \mathbb{R}$ or $\mathbb{C}$ which is also the unique solution of (1.2.24)
(1.2.31) in the closure of the span of

$S = \{1,\ e^{-2i\omega_j t},\ e^{2i\omega_j t}\ |\ t \in [0,T],\ j \in \mathbb{N}\}$ and has a representation

$$u^\infty(t) = \sum_{j=0}^\infty c_j\ y_j(t)$$

where $(y_j)_{j \in \mathbb{N}}$ is any sequence in the closure of the span of S with

$$\int_0^T y_k(t)\ z_j(t)\ dt = \delta_{kj} \quad \text{for all } j,\ k \in \mathbb{N} \cup \{0\},$$

$(z_j)_{j \in \mathbb{N}}$ being defined by (1.2.25) and $z_0 \equiv 1$.

In addition to (1.2.24) and (1.2.31) we consider the equation

$$\int_0^T t\ u(t)\ dt = c_{-1} \tag{1.2.32}$$

for some $c_{-1} \in \mathbb{R}$ or $\mathbb{C}$.

Then we have the

Lemma 1.2.13: Under the assumption of Lemma 1.2.10 the sequence $\{t, 1, z_j(t)\ |\ t \in [0,T],\ j \in \mathbb{N}\}$ is minimal where $(z_j)_{j \in \mathbb{N}}$ is defined by (1.2.25).

Proof: By the Corollary of Lemma 1.2.10 we only have to prove that $z_{-1}(t) = t$, $t \in [0,T]$, is not in the closure V of the span of $S = \{1, z_j\ |\ j \in \mathbb{N}\}$. Let us assume that this were not true, i.e., $z_{-1} \in V$. Then it follows that

$$\lim_{N \to \infty} |z_{-1} - v_N|_{L^2(0,T)} = 0 \quad \text{where } v_N = \sum_{j=0}^N a_j\ z_j,\ a_j^N \in \mathbb{R} \text{ or } \mathbb{C}.$$

Then for every $\varphi \in C^1[0,T]$ with $\varphi(0) = \varphi(T) = 0$ we conclude that

$$0 = \lim_{N \to \infty} \int_0^T (z_{-1}(t) - v_N(t))\varphi'(t)\ dt$$

$$= \lim_{N \to \infty} -\int_0^T (z_0(t) - v_N'(t))\ \varphi(t)\ dt$$

$$= -\int_0^T (z_0(t) - v(t))\varphi(t)\ dt$$

where $\lim_{N \to \infty} |v - v_N'|_{L^2(0,T)} = 0$ and $v_N' \in$ closure of the span of $\{z_j\ |\ j \in \mathbb{N}\}$, hence $z_0 = v \in$ closure of the span of $\{z_j\ |\ j \in \mathbb{N}\}$, a contradiction of the minimality of S which completes the proof.

By Theorem 1.2.6 we therefore have the

Theorem 1.2.14: Under the assumption of Lemma 1.2.10 there is exactly one minimum norm solution $u^\infty \in L^2(0,T)$ of (1.2.24), (1.2.31), (1.2.32) for any choice of $(c_j)_{j \in \mathbb{N}}$ with (1.2.13), c_0, $c_{-1} \in \mathbb{R}$ or $\mathbb{C}$ which is also the unique solution of (1.2.24), (1.2.31), (1.2.32) in the closure of the span of $\{t, 1, z_j(t) \mid t \in [0,T], j \in \mathbb{N}\}$ where $(z_j)_{j \in \mathbb{N}}$ is defined by (1.2.25).

The next step will be to relax the condition (1.2.29), i.e., the assumption of Lemma 1.2.10, to the requirement

$$\liminf_{j \to \infty} \|\omega_j - \omega_{j-1}\| > \frac{\pi}{T} \qquad (1.2.33)$$

which implies that there is some $\varepsilon > 0$ and some $N(\varepsilon) \in \mathbb{N}$ such that

$$\omega_j - \omega_{j-1} \geq \frac{\pi + \varepsilon}{T} \quad \text{for all } j \geq N(\varepsilon). \qquad (1.2.34)$$

This in turn implies (as in the proof of Lemma 1.2.10) that (1.2.30) is satisfied for all $N \geq N(\varepsilon)$ and all $(a_{N(\varepsilon)}, \ldots, a_N) \in \mathbb{R}^{N-N(\varepsilon)+1}$ or $\mathbb{C}^{N-n(\varepsilon)+1}$ with $(z_j)_{j \in \mathbb{N}}$ being defined by (1.2.25) and $z_0 \equiv 1$.

Thus Theorem 1.2.6 could be applied, if we could ensure the minimality of $\{t, 1, z_j(t) \mid t \in [0,T], j \in \mathbb{N}\}$ in $L^2(0,T)$.

For this purpose we first consider the system

$$S = \{1, z_j(t) \mid t \in [0,T], j \in \mathbb{N}\}$$

$$= \{1, e^{-2i\omega_j t}, e^{2i\omega_j t} \mid t \in [0,T], j \in \mathbb{N}\} \qquad (1.2.35)$$

and show that, if S is incomplete in $L^2(0,T)$, then S is minimal in $L^2(0,T)$. Then we give a condition which ensures that S is incomplete, hence minimal, in $L^2(0,T)$ and that the system $S \cup \{t\}$ is also minimal in $L^2(0,T)$.

Theorem 1.2.15: If the system S (1.2.35) is incomplete in $L^2(0,T)$ for some $T > 0$, then S is minimal in $L^2(0,T)$.

Proof: We put $\omega_0 = 0$ and $\omega_{-j} = -\omega_j$. Then the incompleteness of S in $L^2(0,T)$ implies the existence of a function $g \in L^2(0,T)$ not vanishing almost everywhere such that

$$\int_O^T g(t) \; e^{2i\omega_j t} \; dt = O \quad \text{for all } j \epsilon \mathbb{Z}. \qquad (1.2.36)$$

For every $\omega \epsilon \mathbb{R}$ we define

$$J(\omega) = \int_O^T g(t) \; e^{2i\omega t} \; dt.$$

Then (1.2.36) implies

$$J(\omega_j) = O \quad \text{for all } j \epsilon \mathbb{Z}. \qquad (1.2.37)$$

Furthermore, $J(\omega)$ can also be written as

$$J(\omega) = \int_{-T}^T \tilde{g}(s) \; e^{i\omega s} \; ds, \quad \omega \epsilon \mathbb{R},$$

where

$$\tilde{g}(s) = \tfrac{1}{2}g(\tfrac{1}{2}(s+T))e^{i\omega T}, \quad s \epsilon [-T,T]$$

Obviously $J = J(\omega)$ is an entire function on $\mathbb{R}$ and has only zeros of finite multiplicity. If ω_k is a zero of J of multiplicity $m_k \geq 1$, then we define

$$J_k(\omega) = \frac{J(\omega)}{(\omega-\omega_k)^{m_k-1} \; J^{(m_k)}(\omega_k)}, \quad \omega \epsilon \mathbb{R}.$$

This implies

$$J_k(\omega_j) = \delta_{kj} \quad \text{for all } k, \; j \epsilon \mathbb{N}.$$

Since every $J_k = J_k(\omega)$ is in $L^2(-\infty,\infty)$ and of exponential type, it can be represented, by a theorem of Paley and Wiener, in the form

$$J_k(\omega) = \int_{-T}^T \tilde{y}_k(s) \; e^{i\omega s} \; ds, \quad \omega \epsilon \mathbb{R},$$

for some $\tilde{y}_k \epsilon L^2(-T,T)$. So, if we define

$$y_k(t) = 2\tilde{y}_k(2t-T) \; e^{-i\omega_k T} \quad \text{for } t \epsilon [O,T],$$

we obtain

$$J_k(\omega_j) = \int_O^T y_k(t) \; e^{2i\omega_j t} \; dt = \delta_{kj} \quad \text{for all } k, \; j \epsilon \mathbb{Z},$$

i.e., $(y_k)_{k\epsilon Z}$ is a biorthonormal sequence of S in $L^2(0,T)$
and hence S is minimal by Theorem 1.2.5. This completes
the proof.

In order to formulate the above mentioned condition which
implies the incompleteness of S (1.2.35) and the minimality
of $Su\{t\}$ in $L^2(0,T)$ we introduce a density D of the sequence
$(\omega_j)_{j\epsilon Nu\{0\}}$ by defining

$$D = \lim_{y\to\infty} \sup \left(\lim_{x\to\infty} \sup \frac{d(x+y) - d(x)}{y} \right) \qquad (1.2.38a)$$

where, for every x>0,

$$d(x) = \text{number of } 2\omega_j < x. \qquad (1.2.38b)$$

Then we have (see [17])

Theorem 1.2.16: Let

$$D < \frac{T}{2\pi}. \qquad (1.2.39)$$

Then there exists an entire function H on $(-\infty, +\infty)$ with $H(0) \neq 0$
such that

$$G(\omega) = \omega H(\omega) \prod_{j=1}^{\infty} (1 - \frac{4\omega^2}{4\omega_j^2})$$

is in $L^2(-\infty, \infty)$ and has a representation of the form

$$G(\omega) = \int_0^T g(t) e^{2i\omega t} dt, \quad \omega\epsilon(-\infty, +\infty), \qquad (1.2.40)$$

with some $g\epsilon L^2(0,T)$ which does not vanish almost everywhere.

As a consequence of Theorem 1.2.16 we have the

Theorem 1.2.17: If (1.2.39) is satisfied, then the system S (1.2.35)
is incomplete, hence minimal, in $L^2(0,T)$ and the system $Su\{t\}$ is
also minimal in $L^2(0,T)$.

Proof: The first assertion follows immediately from

$$G(\omega_j) = 0 \quad \text{for all } j\epsilon Z$$

and the representation (1.2.40) by virtue of Theorem 1.2.15.

From

$$t = \lim_{\omega \to 0} \frac{e^{2i\omega t} - e^{-2i\omega t}}{4i\omega}$$

it follows that

$$\int_0^T g(t) t \, dt = \lim_{\omega \to 0} \int_0^T g(t) \frac{e^{2i\omega t} - e^{-2i\omega t}}{4i\omega} \, dt$$

$$= \lim_{\omega \to 0} \frac{G(\omega) - G(-\omega)}{4i\omega} = -\frac{i}{4} \frac{dG}{d(\omega)}(0) = -\frac{i}{4} H(0) \neq 0$$

which in connection with

$$\int_0^T g(t) e^{2i\omega_j t} \, dt = G(\omega_j) = 0 \quad \text{for all } j \in \mathbb{Z}$$

and the minimality of S (1.2.35) shows that $S \cup \{t\}$ is also minimal in $L^2(0,T)$.

Summarizing we obtain, by Theorem 1.2.6, the

Theorem 1.2.18: Let (1.2.33) and (1.2.39) be satisfied. For every sequence $(c_j)_{j \in \mathbb{N}}$ with (1.2.13) and every choice of c_{-1}, $c_0 \in \mathbb{R}$ or $\mathbb{C}$ we then have the following two assertions:

a) There is exactly one minimum norm solution $u^\infty \in L^2(0,T)$ of (1.2.24) which is also the unique solution of (1.2.24) in the closure of the span of

$S = \{e^{-2i\omega_j t}, e^{2i\omega_j t} \mid t \in [0,T], j \in \mathbb{N}\}$ and has a representation

$$u^\infty(t) = \sum_{j=1}^\infty c_j y_j(t)$$

where $(y_j)_{j \in \mathbb{N}}$ is any sequence in the closure of the span of S with

$$\int_0^T y_k(t) z_j(t) \, dt = \delta_{kj} \quad \text{for all } j, k \in \mathbb{N}$$

and $(z_j)_{j \in \mathbb{N}}$ being defined by (1.2.25).

b) There is exactly one minimum norm solution $u^\infty \in L^2(0,T)$ of (1.2.24), (1.2.31), (1.2.32) which is also the unique solution of (1.2.24), (1.2.31), (1.2.32) in the closure of the span of S (1.2.35)$\cup\{t\}$, $t \in [0,T]$, and has a representation

$$u^\infty(t) = \sum_{j=-1}^{\infty} c_j \, y_j(t)$$

where $(y_j)_{j \in \mathbb{N} \cup \{0\} \cup \{-1\}}$ is any biorthonormal sequence of S (1.2.35) in the closure of the span of $S \cup \{t\}$.

We shall see in Section 1.2.3.3 that for control problems which are governed by differential equations of second order the following condition can be verified:

$$\lim_{j \to \infty} \frac{j}{2\omega_j} = \Gamma > 0. \tag{1.2.41}$$

If this is the case, then we can prove

$$D \leq \Gamma \tag{1.2.42}$$

where D is the density defined by (1.2.38).

In order to show this we choose $\varepsilon > 0$ arbitrarily and have

$$|j - 2\omega_j \Gamma| \leq \varepsilon \quad \text{for all } j \geq j(\varepsilon).$$

Let $j = d(x)$ be the largest index such that $2\omega_j < x$. Then

$$d(x) \leq 2\omega_j \Gamma + \varepsilon \leq x\Gamma + \varepsilon,$$

if x is sufficiently large. Further it follows that $x \leq 2\omega_{j+1}$ and therefore

$$x\Gamma - \varepsilon \leq 2\omega_{j+1} \Gamma - \varepsilon \leq j + 1 = d(x) + 1$$

or

$$- d(x) \leq - x\Gamma + \varepsilon + 1,$$

if x is sufficiently large. So we have

$$\frac{d(x+y) - d(x)}{y} \leq \frac{(x+y)\Gamma + \varepsilon - x\Gamma + \varepsilon + 1}{y} = \Gamma + \frac{2\varepsilon + 1}{y},$$

if x and y are sufficiently large, which implies (1.2.42). Hence

$$\Gamma < \frac{T}{2\pi} \tag{1.2.43}$$

is sufficient for (1.2.39), if (1.2.41) is satisfied.

For the above control problems which guarantee (1.2.41) to
be true one can also show that

$$\lim_{j \to \infty} \inf \left(\omega_j - \omega_{j-1} \right) = \frac{1}{2\Gamma} \tag{1.2.44}$$

so that (1.2.43) also implies (1.2.33) and thus ensures the
assumption of Theorem 1.2.18.

Next we assume (1.2.41) to be satisfied but (1.2.43) to be
violated. Then we have to consider two cases:

a) $\Gamma > \frac{T}{2\pi}$. $\tag{1.2.45}$

If (1.2.45) holds, then by [19], footnote on page 3, it follows
that the system S (1.2.35) is complete and stays complete, if
finitely many of its elements are removed. As a consequence there
is, in general, no solution of (1.2.24) because every element of
S is in the closure of the span of the others. So, if $u \in L^2(0,T)$
is a solution of (1.2.24), i.e.,

$$\int_0^T u(t) \, z_j(t) \, dt = c_j \quad \text{for all } j \in \mathbb{N},$$

then

$$z_j = \lim_{N \to \infty} \sum_{\substack{k=1 \\ k \neq j}}^{N} a_k^N z_k ,$$

hence

$$c_j = \lim_{N \to \infty} \sum_{\substack{k=1 \\ k \neq j}}^{N} a_k^N c_k ,$$

i.e., c_j is in the closure of the span of $\{c_k \mid k \neq j\}$, a requirement
which in general is not satisfied.

b) $\Gamma = \frac{T}{2\pi}$. $\tag{1.2.46}$

In this case we can make use of a result by Kadec [4] which can
be stated in the following form (see also [17]): If there exists
a constant K>0 such that

$$\sup_{j \in \mathbb{Z}} |2\omega_j - \frac{j}{K}| < \frac{1}{4k}, \tag{1.2.47}$$

then the system S (1.2.35) is a Riesz basis in $L^2(0,2K\pi)$, i.e.,
every $u\in L^2(0,2K\pi)$ has a unique representation of the form

$$u(t) = a_0 + \sum_{k=1}^{\infty} a_{2k-1}\ e^{-2i\omega_k t} + a_{2k}\ e^{2i\omega_k t} \tag{1.2.48}$$

such that

$$m \sum_{k=0}^{\infty} |a_k|^2 \le |u|^2_{L^2(0,K\pi)} \le M \sum_{k=0}^{\infty} |a_k|^2 \tag{1.2.49}$$

where m, M are two constants with $0<m\le M$ which are independent
of u.

Let us assume that (1.2.47) is satisfied for $K = \frac{T}{2\pi}$ which implies
(1.2.41) for $\Gamma = \frac{T}{2\pi}$. Then S (1.2.35) is a Riesz basis in $L^2(0,T)$
and the assertion of Lemma 1.2.10 is satisfied so that the Theorems
1.2.11 and 1.2.12 can be applied. Since $u(t) = t$, $t\in[0,T]$, has a
unique representation (1.2.48), it follows that (1.2.24), (1.2.31),
(1.2.32) has a unique minimum norm solution $u^\infty \in L^2(0,T)$, if and
only if

$$c_{-1} = c_0\ a_0 + \sum_{j=1}^{\infty} c_j\ a_j . \tag{1.2.50}$$

1.2.3. Application to One-Dimensional Vibrations.

1.2.3.1. Distributed Control.

We come back to the situation of Section 1.1.1 where the problem of null-controllability of a one-dimensional vibrating system by distributed control was reduced to the solvability of (1.1.23) for all $j \in N$. Let us define

$$\omega_j = \frac{1}{2} \sqrt{\lambda_j} \quad \text{for all } j \in \mathbb{N} \tag{1.2.51}$$

$$c_{2j-1} = c_j^1 + ic_j^2 \text{ and } c_{2j} = c_j^1 - ic_j^2, \ j \in \mathbb{N}, \tag{1.2.52}$$

where c_j^1 and c_j^2 are given by (1.1.23).

Then the system (1.1.23) for all $j \in \mathbb{N}$ is equivalent to the system (1.2.24) in the following sense: Every solution $u \in L^2(0,T)$ of (1.1.23) for all $j \in \mathbb{N}$ is a real solution of (1.2.24). If $u \in L^2(0,T)$ is a solution of (1.2.24) then the conjugate complex $\bar{u}$ of u is also a solution of (1.2.24) and $\frac{1}{2}(u + \bar{u})$ is a real solution of (1.2.24) which is also a solution of (1.1.23) for all $j \in \mathbb{N}$. Every unique minimum norm solution of (1.2.24) is necessarily real and therefore also a minimum norm solution of (1.1.23) for all $j \in \mathbb{N}$. Conversely every such is a minimum norm solution of (1.2.24).

From the properties of the sequence $(\lambda_j)_{j \in \mathbb{N}}$ (see Section 1.1.1) it follows that the sequence $(\omega_j)_{j \in \mathbb{N}}$ has the properties as being required at the beginning of Section 1.2.2.

Theorem 1.2.11 and its Corollary now lead to

Theorem 1.2.19: Let $\lambda_0 = 0$ and assume that, for a given $T>0$, there is some $\delta > 0$ such that

$$\sqrt{\lambda_j} - \sqrt{\lambda_{j-1}} \geq \frac{2\pi + \delta}{T} \quad \text{for all } j \in \mathbb{N}. \tag{1.2.53}$$

Then, for every sequence $(c_j)_{j \in \mathbb{N}}$ defined by (1.2.52), (1.1.23) such that (1.1.27) is satisfied, there is exactly one minimum norm solution $u^\infty \in L^2(0,T)$ of (1.1.23) for all $j \in \mathbb{N}$ whose norm can be estimated by

$$\|u^\infty\|_{L^2(0,T)} \leq \frac{2A(\varepsilon)^{1/2}}{T} \left(\sum_{j=1}^{\infty} (c_j^1)^2 + (c_j^2)^2 \right)^{1/2} \tag{1.2.54}$$

where $\varepsilon = \frac{1}{2}\delta$ and $A(\varepsilon)$ is given by (1.2.28).

From Theorem 1.2.18a) we deduce

Theorem 1.2.20: Let

$$\liminf_{j\to\infty} \sqrt{\overline{\lambda}_j} - \sqrt{\overline{\lambda}_{j-1}} > \frac{4\pi}{T} \tag{1.2.55}$$

and let

$$\limsup_{y\to\infty} \limsup_{x\to\infty} \frac{d(x+y) - d(x)}{y} < \frac{T}{2\pi} \tag{1.2.56}$$

where, for every x>0,

$$d(x) = \text{number of } \sqrt{\overline{\lambda}_j} < x.$$

Then, for every sequence $(c_j)_{j\in\mathbb{N}}$ as in Theorem 1.2.19, there is exactly minimum norm solution $u^\infty \in L^2(0,T)$ of (1.1.23) for all $j\in\mathbb{N}$.

In both Theorems u^∞ is also the unique solution of (1.1.23) for all $j\in\mathbb{N}$ in the closure of the span of S (1.1.24) in $L^2(0,T)$.

1.2.3.2. Boundary Control.

We pick up the situation of Section 1.1.2 where the problem of null-controllability of a one-dimensional vibrating medium by boundary control was reduced to the solvability of (1.2.23) for all $j\in\mathbb{N}$, (1.1.44), and (1.1.45).

Again we define ω_j and c_j for $j\in\mathbb{N}$ by (1.2.51) and (1.2.52), respectively. In addition we put

$$c_{-1} = c_0 = 0. \tag{1.2.57}$$

Then the system (1.1.23) for all $j\in\mathbb{N}$, (1.1.44), and (1.1.45) is equivalent to (1.2.24), (1.2.31), (1.2.32) in the same sense as (1.1.23) for all $j\in\mathbb{N}$ is equivalent to (1.2.24) being explained at the beginning of Section 1.2.3.1.

Theorem 1.2.14 now leads to

Theorem 1.2.21: Let $\lambda_0 = 0$ and assume that, for a given T>0, there is some $\delta>0$ such that (1.2.53) holds. Then, for every sequence $(c_j)_{j\in\mathbb{N}\cup\{0,-1\}}$ defined by (1.2.57), (1.2.52), (1.1.23) such that (1.1.27) is satisfied, there is exactly one minimum norm solution $u^\infty \in L^2(0,T)$ of (1.1.23) for all $j\in\mathbb{N}$, (1.1.44), and (1.1.45) which is also the unique solution in the closure of the span of S (1.1.24)∪{1,t} in $L^2(0,T)$.

From Theorem 1.2.18b) we deduce

<u>Theorem 1.2.22:</u> Let (1.2.55) and (1.2.56) be satisfied. Then, for every sequence $(c_j)_{j \in \mathbb{N} \cup \{0,-1\}}$ as in Theorem 1.2.21, there is exactly one minimum norm solution $u^\infty \in L^2(0,T)$ of (1.1.23) for all $j \in \mathbb{N}$, (1.1.44), and (1.1.45) which is also the unique solution in the closure of the span of $S(1.1.24) \cup \{1,t\}$ in $L^2(0,T)$.

<u>1.2.3.3 Special Cases.</u>
<u>a) The Vibrating Beam.</u>

We again consider vibrations of a beam as in Sections 1.1.2 and 1.1.4. But here we allow for several boundary conditions of the operator $Lz = z^{(4)}$ on $H^4(0,1)$ which we summarize as

$$z^{(j_1)}(0) = z^{(j_2)}(0) = 0,$$
$$z^{(j_3)}(1) = z^{(j_4)}(1) = 0 \qquad (1.2.58)$$

where the quadruple (j_1, j_2, j_3, j_4) may be chosen from the following table

case	1	2	3	4	5	6
j_1	0	0	0	0	0	0
j_2	1	1	1	1	2	2
j_3	0	0	1	2	0	1
j_4	1	2	3	3	2	3

$$(1.2.59)$$

In all 6 cases of (1.2.59) it can be shown that L is self-adjoint and positive definite on

$$D_L = \{z \in H^4(0,1) \mid z \text{ satisfies } (1.2.58)\}$$

and the eigenvalues of L are given by

$$\lambda_j = [(j-\sigma)\pi + \varepsilon_j]^4, \quad j \in \mathbb{N}, \qquad (1.2.60)$$

with

case	1	2	3	4	5	6
σ	$-\frac{1}{2}$	$-\frac{1}{4}$	$\frac{1}{4}$	$\frac{1}{2}$	0	$\frac{1}{2}$

$$(1.2.61)$$

$$|\varepsilon_j| < \frac{\pi}{4} \quad \text{for all } j \in \mathbb{N}$$

$$\varepsilon_j \to 0 \quad \text{as } j \to \infty$$

$$(1.2.62)$$

(see, for instance[*]).

From (1.2.60) it follows that

$$\liminf_{j \to \infty} \sqrt{\lambda}_j - \sqrt{\lambda}_{j-1} = \infty$$

$$(1.2.63)$$

so that (1.2.55) is true for every T>0.

But we also have

$$\limsup_{y \to \infty} \quad \limsup_{x \to \infty} \frac{d(x+y) - d(x)}{y} = 0$$

$$(1.2.64)$$

where, for every x>0,

$$d(x) = \max\{j \in \mathbb{N} | \sqrt{\lambda}_j < x\}$$

so that (1.2.56) is also satisfied for every T>0.

In order to show (1.2.64) we first observe that $\sqrt{\lambda}_j < x$ implies

$$(j-\sigma)\pi + \varepsilon_j < \sqrt{x}$$

or

$$j < \frac{1}{\pi} \sqrt{x} + \sigma - \frac{\varepsilon_j}{\pi} \le \frac{1}{\pi}\sqrt{x} + \frac{3}{4},$$

hence

$$d(x) \le [\frac{1}{\pi}\sqrt{x} + \frac{7}{4}] \quad \text{for all } x > 0$$

where $[\alpha]$ denotes the largest $k \in \mathbb{N}$ with $k \le \alpha$, $\alpha \ge 0$. For a given x>0 we put

$$j = [\frac{1}{\pi}\sqrt{x} + \sigma - \frac{1}{4}].$$

Then

$$((j-\sigma)\pi + \frac{\pi}{4})^2 \le x \implies \sqrt{\lambda}_j < x,$$

hence

$$[\frac{1}{\pi}\sqrt{x} + \sigma - \frac{1}{4}] \le d(x)$$

---

[*] Coddington, E.A. and Levinson, N.: Theory of Ordinary Differential Equations. McGraw-Hill: New York - Toronto - London 1955)

for x>0 sufficiently large. As a result we obtain for x>0 sufficiently large and y>0

$$0 \leq \frac{d(x+y) - d(x)}{y} \leq \frac{\frac{1}{\pi}\sqrt{x+y} + \frac{7}{4} - \frac{1}{\pi}\sqrt{x} - \sigma + \frac{5}{4}}{y}$$

$$= \frac{1}{\pi(\sqrt{x+y} + \sqrt{x})} + \frac{3 - \sigma}{y}$$

which completes the proof of (1.2.64).

As a consequence the assertions of Theorems 1.2.20 and 1.2.22 hold true for every T>0. Thus null-controllability is possible with distributed as well as boundary control for every time T>0 and initial states (y_0, y_1) such that the corresponding Fourier coefficients a_j, b_j (1.1.16) satisfy the condition of (1.1.27).

The situation is quite different in the case of

b) The Vibrating String.

As in the Sections 1.1.2 and 1.1.4 we consider vibrations of a string. But here we allow for the following boundary conditions of the operator $Lz = - z''$ on $H^2(0,1)$:

$$z^{(j_1)}(0) = z^{(j_2)}(1) = 0 \tag{1.2.65}$$

where the pair (j_1, j_2) may be chosen from the following table

case	1	2	3	
j_1	0	0	1	(1.2.66)
j_2	0	1	0	

In all three cases of (1.2.66) L is symmetric and positive definite on

$$D_L = \{z \in H^2(0,1) \mid z \text{ satisfies } (1.2.65)\}$$

and the eigenvalues of L are given by

$$\lambda_j = (j\pi)^2 \text{ in case 1}$$
$$\lambda_j = [(j - \frac{1}{2})\pi]^2 \text{ in the cases 2 and 3, } j \in \mathbb{N}. \tag{1.2.67}$$

Thus we have

$$\lim_{j\to\infty} \frac{j}{2\omega_j} = \lim_{j\to\infty} \frac{j}{\sqrt{\lambda_j}} = \frac{1}{\pi}$$

in all three cases, i.e., the condition (1.2.41) is satisfied for $\Gamma = \frac{1}{\pi}$. Hence T>2 is equivalent with (1.2.43) which implies (1.2.39).

We also have

$$\liminf_{j\to\infty} \omega_j - \omega_{j-1} = \liminf_{j\to\infty} \frac{1}{2}\sqrt{\lambda}_j - \frac{1}{2}\sqrt{\lambda}_{j-1} = \frac{\pi}{2},$$

i.e., the condition (1.2.44) is also satisfied for $\Gamma = \frac{1}{\pi}$ and implies (1.2.33). So the assumptions of Theorem 1.2.18 hold and thus the assertion of Theorem 1.2.20 is true for every T>2. This can also be shown by verifying (1.2.55) and (1.2.56). The first of these conditions has been deduced above already. In order to show (1.2.56) we first mention that

$$d(x) = [\frac{x}{\pi}] \quad \text{for all } x > 0 \text{ with } x \neq j\pi, \ j\epsilon \ N,$$

in case 1 and

$$d(x) = [\frac{x}{\pi} + \frac{1}{2}] \quad \text{for all } x > 0 \text{ with } x \neq (j - \frac{1}{2})\pi, \ j\epsilon \ N,$$

in the cases 2 and 3. This implies

$$\frac{d(x+y) - d(x)}{y} \leq \frac{1}{\pi} \quad \text{for all } x, y > 0$$

in all three cases from which (1.2.56) follows, if T>2.

If T<2, then (1.2.45) holds for $\Gamma = \frac{1}{\pi}$ and in general there is no solution of (1.2.24) as being shown in Section 1.2.2. In turn (1.1.23) for all $j\epsilon \ N$ has also no solution in general.

For T = 2 the case 1 has been studied in Section 1.1.4 already by elementary means. The moment problems corresponding to the other two cases can be transformed to moment problems of the same form as in case 1.

1.2.4. Bibliographical Remarks and References

The idea of applying moment theory to problems of controllability
is not new. One of the first to observe this possibility was
Butkovskiy who devoted chapter 3 of his book [2] to this topic.
He also gives references there of his own contributions and of
some earlier work of Kraskovskii in this direction. His representation of the moment theory is based on the monograph [1] by
Akhieser and Krein in which Krein deals with the so called
"L-Problem (of Moments) in an Abstract Linear Normed Space". The
theory developed in Section 1.2.1 can be considered as a supplement
and partial refinement of the approach by Butkovskiy and Krein in
the case of Hilbert space problems. It is based on results in the
paper [5] of Korobeinik which originate from earlier work of Lewin
in [9] and [10]. A systematic use of this theory was made in [7]
and [8].

As an application of infinite moment theory Butkovskiy deals with
the vibrating string being symmetrically controlled at both ends
in the smallest possible time and under right end control within
any time being greater than or equal to the least possible time
of control. The requirement that the string stay in rest when
the control that achieves the state of rest is turned off, however,
is neglected. Butkovskiy gives explicit representations of least
norm controls which in the case of smallest possible time corre-
spond to the results in Section 1.1.4 when the compatibility
condition (1.1.50') is met.

In [13] Parks also advocates the application of moment theory to
automatic control. Among others he treats the problem of null-
controllability of a vibrating string by steering at the right
end. He also neglects the requirement that the system stay in rest
when the control is turned off.

The theory of trigonometric moment problems being developed in
Section 1.2.2 rests on the pioneering work [17] by Russell
(which was also pointed out in [7] and [8]). He deals with
problems of distributed control as in Section 1.1.1 for n = 2,
however, also allowing for differential operators L in (1.1.1)
that have zero as an eigenvalue. If this is the case, then the
condition (1.1.5) of null-controllability becomes equivalent to

the moment problem (1.1.23) supplemented by two more equations
of the form

$$\int_0^T u(t)\,dt = c_0^1 \quad \text{and} \quad \int_0^T t\,u(t)\,dt = c_0^2.$$

So one is in the same situation as with boundary control where
the additional equations (1.1.44) and (1.1.45) come in. If all
the eigenvalues of L in (1.1.1) are positive, then (1.1.5) turns
out to be equivalent to (1.1.23). This is the case we here
restricted to.

In [17] Russell also makes use of moment theory in a Hilbert
space but does not give it a systematic account. His main tools
are results of Ingham in [3], Redheffer in [15], and Schwartz in
[19]. We follow his lines by first proving Ingham's result as
Theorem 1.2.9 which finally leads to Theorem 1.2.14. In order to
obtain controllability results for arbitrarily small time
intervals the inequality (1.2.29) which is the main assumption
in Theorem 1.2.14 has to be relaxed to the inequality (1.2.33).
In the further development the Theorem 1.2.15 which is due to
Schwartz and Theorem 1.2.16 taken from Russell's paper [17] play
a decisive role. Russell derived this result from the paper [15]
by Redheffer.

If the two conditions (1.2.33) and (1.2.39) which are the main
assumptions of Theorem 1.2.18 were replaced by

$$\omega_j - \omega_{j-1} \to \infty \quad \text{as} \quad j \to \infty, \tag{1.2.68}$$

then, by virtue of Theorem 1.4 in [14], the first part of Theorem
1.2.18 could also be proved, since (1.2.68) ensures the minimality
of $\{e^{zi\omega_j t} \mid j \in Z, \ t \in [0,T]\}$ (where $\omega_{-j} = -\omega_j$) for every $T > 0$.

In [18] Russell continued the investigations of [17] by reversing
the argument in giving a direct proof for controllability and
deriving conclusions for the corresponding moment problem. Instead
of distributed control he considers boundary control at the right
end. A detailed investigation is made for the "critical time" which
corresponds to the case $T = 2\pi\Gamma$ (see (1.2.46)) with Γ defined by

(1.2.41). In [12] Parks treats the much simpler case of the
homogeneous vibrating string with boundary control of the
deviation at the right end. Here the critical time is $T = 2$
and the optimal control which transfers the initial state
(y_0, y_1) to rest can be explicitly expressed in terms of y_0
and y_1. In [6] it is shown that this optimal control minimizes
the vibration energy for each $t \in [0,2]$.

Referring to the paper [16] by Russell the problem of boundary
control of a vibrating string under Dirichlet boundary conditions is also treated by Malanowski in [11]. But instead of
the displacement $y(x,t)$ the solution $y^*(x,t)$ of $(1.1.1^*)$,
$(1.1.3^*)$, $(1.1.4^*)$ with $u = - v'' \in L^2[0,T]$ is controlled so that
basicly a problem of distributed control as in Section 1.1.1 is
solved. The solution is given without moment theory in terms of
an explicit formula for a time-minimal norm-bounded control.

References

[1] Akhieser, N.J., and Krein, M.: Some Questions in the Theory of
     Moments. Providence: American Mathematical Society 1962 =
     Translations of Mathematical Monographs, No 2.

[2] Butkovskiy, A.G.: Theory of Optimal Control of Distributed
     Parameter Systems. New York - London - Amsterdam: Elsevier 1969.

[3] Ingham, A.E.: Some Trigonometrical Inequalities with Applications
     to the Theory of Series. Math. Z. 41 (1936), 367-379.

[4] Kadec, M.J.: The Exact Value of the Paley-Wiener Constant.
     Sov. Math. 5, No. 1 (1964), 559-561.

[5] Korobeinik, J.F.: The Moment Problem, Interpolation and
     Basicity. Math. USSR, Isvestija 13 (1979), 277-306.

[6] Krabs, W.: Über die einseitige Randsteuerung einer schwingenden
     Saite in einem Zustand minimaler Energie. Computing 7 (1977),
     351-359.

[7] Krabs, W.: On Boundary Controllability of One-Dimensional
     Vibrating Systems. Math. Meth. in the Appl. Sci. 1 (1979),
     322-345.

[8] Krabs, W.: Optimal Control of Processes Governed by Partial
     Differential Equations. Part II: Vibrations. ZOR 26 (1982),
     63-86.

[9]   Lewin, S.: Über einige mit der Konvergenz im Mittel verbun-
      denen Eigenschaften von Funktionenfolgen. Math. Z. 32 (1930),
      491-511.

[10] Lewin,S.: Integralgleichungen und Funktionenräume. Mat. Sb.
      39, No. 4 (1932), 3-72.

[11] Malanowski, K.: On Time-Optimal Control of a Vibrating String
      (Polish). Archiwum Automatyki i Telemechaniki XIV (1969),
      33-45.

[12] Parks, P.C.: On How to Shake a Piece of String to a Standstill.
      In: Recent Mathematical Developments in Control. Ed. by D.J.
      Bell. London - New York: Academic Press 1973, 267-287.

[13] Parks, P.C.: Applications of the Theory of Moments in Auto-
      matic Control. Intern. J. of Syst. Sc. 7 (1976), 177-189.

[14] Quinn, Ph.J.: The Optimal Control of Linear Distributed
      Parameter Systems. Ph.D. Thesis, Wisconsin 1969.

[15] Redheffer, R.M.: Remarks on Incompleteness of $\{e^{i\lambda_n x}\}$, Non-
      Averaging Sets, and Entire Functions. Proc. Amer. Math. Soc.
      2 (1951), 365-369.

[16] Russell, D.L.: Optimal Regulation of Linear Symmetric Hyper-
      bolic Systems with Finite Dimensional Controls. SIAM J. on
      Control 4 (1966), 276-294.

[17] Russell, D.L.: Non-Harmonic Fourier Series in Control Theory
      of Distributed Parameter Systems. J. Math. Anal. Appl. 18
      (1967), 542-560.

[18] Russell, D.L.: Control Theory of Hyperbolic Equations Related
      to Certain Questions in Harmonic Analysis and Spectral Theory.
      J. Math. Anal. Appl. 40 (1972), 336-368.

[19] Schwartz, L.: Etude des Sommes D'exponetielles. Paris:
      Hermann 1959.

1.3. On the Solvability of Linear Operator Equations.

1.3.1. Exact Solvability.

In order to solve the problem of null-controllability as formulated
in Sections 1.1.1 and 1.1.3 one can also adopt the following
point of view. Let X and Y be two normed linear spaces over the
real or the complex numbers and let S be a linear mapping from X
into Y. Question: Under which conditions does there exist, for
any choice of $y \epsilon Y$, some $x \epsilon X$ such that

$$S(x) = y, \tag{1.3.1}$$

in other words, under which conditions does S map X onto Y?

One can also investigate the solvability of the operator equation
(1.3.1) for an individual element $y \epsilon Y$ which will not be the concern
in this section. We will not always assume that the mapping (or the
operator) S is defined on all of X but on a linear subspace of X
denoted by $D(S)$ and called the domain of S. In this case we ask
whether S maps $D(S)$ onto Y.

A basic property that we will frequently require is the <u>closedness</u>
of S which means that its graph

$$G(S) = \{ (x, S(x)) \mid x \epsilon D(S) \} \tag{1.3.2}$$

be closed in X×Y equipped with the norm

$$| (x,y) |_{X \times Y} = \max(|x|_X, \ |y|_Y), \ x \epsilon X, \ y \epsilon Y.$$

Also the <u>adjoint operator</u> S^* of $S : D(S) \rightarrow Y$ will play a fundamental
role. In order to define S^* we assume $D(S)$ to be dense in X, i.e.,
$\overline{D(S)} = X$, where $\overline{A}$ denotes the closure of a subset of a normed linear
space. Let X^* and Y^* denote the dual space of X and Y, respectively.
Then the domain of S^* is defined by

$$D(S^*) = \{ y^* \epsilon Y^* \mid y^* S : D(S) \rightarrow \mathbb{R} \text{ or } C \text{ is continuous} \} \tag{1.3.3}$$

and $S^*: D(S^*) \rightarrow X^*$ is defined by

$$S^*(y^*)(x) = y^* S(x) \tag{1.3.4}$$
$$\text{for all } x \epsilon D(S) \text{ and } y^* \epsilon D(S^*).$$

Since $D(S)$ is dense in X, every $S^*(y^*)$ can be uniquely extended to
a continuous linear functional on all of X which justifies the
notation $S^*: D(S^*) \rightarrow X^*$. Before we give an answer to the above question

we will prove two auxiliary results which are useful for typical situations in the application of the theoretical results to applications.

Lemma 1.3.1: Let $S : X \to Y$ be a linear mapping with $D(S) = X$ such that X is complete and the range $R(S) = S(D(s)) = S(X)$ of S is contained in a linear subspace Y_1 of Y whose norm $|.|_{Y_1}$ satisfies, for some $\lambda > 0$,

$$|y|_Y \leq \lambda |y|_{Y_1} \quad \text{for all } y \in Y_1 \tag{1.3.5}$$

and which is $|.|_{Y_1}$-complete. If S is continuous, i.e.,

$$|S(x)|_Y \leq \alpha |x|_X \quad \text{for all } x \in X$$

and some $\alpha > 0$, then $S : X \to Y_1$ is closed with respect to the norm $|.|_{Y_1}$ in Y_1.

Proof: We consider a sequence (x_n) in X which converges to some $x \in X$ and for which the image sequence $(S(x_n))$ converges to some $y \in Y_1$. Then we have to show that $S(x) = y$. In order to see this we observe that $x_n \to x$ implies

$$\lim_{n \to \infty} |S(x_n) - S(x)|_Y = 0,$$

by virtue of the continuity of S, and that

$$|S(x_n) - y|_{Y_1} \to 0 \text{ implies } |S(x_n) - y|_Y \to 0$$

as a consequence of (1.3.5). The assertion $S(x) = y$ then follows from the uniqueness of the limit of a converging sequence.

Lemma 1.3.2: Let $S : X \to Y$ be a linear mapping whose domain $D(S)$ is dense in X and let the domain $D(S^*)$ of the adjoint operator S^* of S be total, i.e., $y^*(y) = 0$ for all $y^* \in D(S^*)$ and some $y \in Y$ implies that y is the zero element of Y. Then S can be extended to a closed linear operator $\tilde{S}$ and $\tilde{S}^* = S^*$.

Proof: Let $\overline{G(S)}$ be the closure of the graph $G(S)$ of S (see (1.3.2)) in $X \times Y$. We first show that

$(\Theta_X, y) \notin \overline{G(S)}$, if $y \neq \Theta_Y$,

where Θ_X and Θ_Y denote the zero elements of X and Y respectively. If this were not true, there were a sequence (x_n) in D(S) with

$$\lim_{n \to \infty} x_n = \Theta_X \text{ and } \lim_{n \to \infty} S(x_n) = y.$$

For every $y^* \in D(S^*)$ we then conclude that

$$y^*(y) = \lim_{n \to \infty} y^* S(x_n) = \lim_{n \to \infty} S^*(y^*)(x_n) = 0,$$

hence $y = \Theta_Y$, a contradiction.

If we define the domain of $\tilde{S}$ by

$$D(\tilde{S}) = \{x | (x,y) \in \overline{G(S)} \text{ for some } y \in Y\}$$

and put $\tilde{S}(x) = y$, then $\tilde{S}$ is well defined and is a closed extension of S to $D(\tilde{S})$.

Let $y^* \in D(\tilde{S}^*)$. Then $y^* \tilde{S}$ is continuous on $D(\tilde{S})$ hence on D(S), and $y^* \in D(S^*)$. Suppose $y^* \in D(S^*)$ and $x \in D(\tilde{S})$. Since $(x, \tilde{S}(x)) \in G(\tilde{S}) = \overline{G(S)}$, there exists a sequence (x_n) in D(S) such that $x_n \to x$ and $S(x_n) \to \tilde{S}(x)$. Hence

$$|y^* \tilde{S}(x)| = \lim_{n \to \infty} |y^* S(x_n)| \leq |S^*(y^*)| \lim_{n \to \infty} |x_n|_X$$

$$= |S^*(y^*)| \ |x|_X.$$

Therefore $y^* \tilde{S}$ is continuous which means $y^* \in D(\tilde{S}^*)$. Consequently, $D(S^*) = D(\tilde{S}^*)$. Since

$$S^*(y^*) = \tilde{S}^*(y^*) \text{ on } D(S), \text{ it follows that } S^* = \tilde{S}^*$$

which completes the proof.

This result has a very useful

Corollary: A linear operator S from a Banach space X into a Banach space Y is continuous, if and only if the domain $D(S^*)$ of its conjugate operator S^* is total.

Proof: If S : X→Y is continuous, then $D(S^*) = Y^*$ is total. If this is the case, then S is closed by Lemma 1.3.2 because it coincides with its closed extension to X and hence continuous by virtue of the closed graph theorem.

The next Theorem now gives a necessary condition for the solvability of (1.3.1) for all $y \in Y$.

Theorem 1.3.3: Let Y be complete and let the linear mapping $S : X \to Y$ have a dense domain $D(S)$ in X. If $R(S) = S(D(S)) = Y$, then S^* has a bounded inverse, i.e., there exists some constant $\lambda > 0$ such that

$$|y^*| \leq \lambda \ |S^*(y^*)| \quad \text{for all } y^* \in D(S^*).\tag{1.3.6}$$

Proof: If the assertion were false, there would exist a sequence (y_n^*) in $D(S^*)$ with

$$|y_n^*| = 1 \quad \text{for all n and } \lim_{n \to \infty} |S^*(y_n^*)| = 0.$$

Put

$$\tilde{y}_n^* = \begin{cases} \dfrac{y_n^*}{|S^*(y_n^*)|^{1/2}}, & \text{if } S^*(y_n^*) \neq 0, \\[2mm] n \cdot y_n^*, & \text{if } S^*(y_n^*) = 0. \end{cases}$$

Then $|\tilde{y}_n^*| \to \infty$ and $|S^*(\tilde{y}_n^*)| \to 0$. Thus, for every $x \in D(S)$, it follows that $\tilde{y}_n^* S(x) = S^*(\tilde{y}_n^*)(x) \to 0$ which implies, in connection with $S(D(S)) = Y$,

$$\sup_n | \tilde{y}_n^*(y)| < \infty \quad \text{for all } y \in Y,$$

hence

$$\sup_n |\tilde{y}_n^*| < \infty \quad \text{by the uniform-boundedness principle}$$

(which requires Y to be complete) which contradicts $|\tilde{y}_n^*| \to \infty$. Therefore the assertion is true.

More important is the converse of Theorem 1.3.3.

Theorem 1.3.4: Let X be complete and again let the linear mapping $S : X \to Y$ have a dense domain $D(S)$ in X. In addition let S be closed. If S^* has a bounded inverse, i.e., if there exists a constant $\lambda > 0$ such that (1.3.6) holds, then $R(S) = Y$.

For the proof we refer to [1].

By combination and specialization of Theorems 1.3.3 and 1.3.4 we obtain the

Theorem 1.3.5: Let X and Y be Banach spaces and let S be a continuous linear mapping from X into Y. Then $S(X) = Y$ holds, if and only if there exists some constant $\lambda > 0$ such that

$$|y^*| \leq \lambda \ |S^*(y^*)| \quad \text{for all } y^* \in Y^*. \tag{1.3.7}$$

1.3.2. Approximate Solvability.

Let $S : X \to Y$ be a linear mapping with domain $D(S)$ and let X and Y be normed linear spaces over $\mathbb{R}$ or $\mathbb{C}$. By approximate solvability of the equation (1.3.1) for a given $y \in Y$ we mean the existence of a sequence (x_n) in $D(S)$ such that

$$\lim_{n \to \infty} S(x_n) = y. \tag{1.3.8}$$

Hence the equation (1.3.1) is approximately solvable for all $y \in Y$, if and only if

$$Y = \overline{R(S)} = \overline{S(D(S))} \tag{1.3.9}$$

where again $\overline{A}$ denotes the closure of a subset A of a normed linear space.

Approximate solvability of (1.3.1) for all $y \in Y$ is governed by the following

Theorem 1.3.6: Let $S : X \to Y$ be a linear mapping whose domain $D(S)$ is dense in X. Then (1.3.9) holds, if and only if the adjoint operator $S^* : D(S^*)$ (1.3.3)$\to X^*$ is injective.

Proof: a) Let (1.3.9) hold. Assume that $S^*(y^*) = \Theta_{X^*}$ = zero element of X^* for some $y^* \in D(S^*)$. Then $y^* S(x) = S^*(y^*)(x)$ for all $x \in D(S)$ and hence $y^*(y) = 0$ for all $y \in Y$, as a consequence of (1.3.9), which implies $y^* = \Theta_{Y^*}$ = zero element of Y^*.

b) Conversely let $S^* : D(S^*) \to X^*$ be injective. Assume (1.3.9) to be false. Then there is some $y \in Y$ with $y \notin \overline{R(S)}$. By a well-known separation theorem for convex sets there exists some $y^* \in Y^*$ with

$$y^*(y) > \sup\{y^*(z) \mid z \in \overline{R(S)}\}.$$

Since $\overline{R(S)}$ is a linear subspace of Y this implies that

$$y^*(z) = O \text{ for all } z \epsilon \overline{R(S)},$$

in particular,

$$s^*(y^*)(x) = y^*S(x) = O \quad \text{for all } x \epsilon D(S)$$

and hence for all $x \epsilon X$ by virtue of the density of $D(S)$ in X.
Thus $s^*(y^*) = \theta_{X^*}$, however, $y^* \neq \theta_{Y^*}$ because of $y^*(y) > 0$. This
contradiction to the injectivity of s^* completes the proof.

1.3.3. Application to Nuclear Operators and Moment Problems in Hilbert Spaces.

Let X and Y be two Hilbert spaces over $\mathbb{R}$ or $\mathbb{C}$. Let $(z_j)_{j \epsilon \mathbb{N}}$ be
a sequence in X such that

$$| \sum_{j=1}^{\infty} a_j z_j |^2_X \leq \mu \sum_{j=1}^{\infty} |a_j|^2 \quad \text{for all } a = (a_j)_{j \epsilon \mathbb{N}} \epsilon l^2 \qquad (1.3.10)$$

and some constant $\mu > 0$ which is independent of a. Finally let
$(y_j)_{j \epsilon \mathbb{N}}$ be a complete orthonormal sequence in Y. First we prove

Lemma 1.3.7: If (1.3.10) holds for some $\mu > 0$, then for every $u \epsilon X$,
the sequence $(<u,z_j>_X)_{j \epsilon \mathbb{N}}$ is in l^2, i.e.,

$$\sum_{j=1}^{\infty} |<u,z_j>_X|^2 < \infty \qquad (1.3.11)$$

Proof: If we define $c_j = <u,z_j>_X$ for all $j \epsilon \mathbb{N}$, then

$$\lambda_N(c) \leq |u|_X \quad \text{for all } N \epsilon \mathbb{N} \qquad (1.2.9)$$

with $\lambda_N(c)$ being defined by (1.2.7).

By Theorem 1.2.1 we have

$$\lambda_N(c)^2 = |u^N|^2_X = < \sum_{j=1}^{N} \xi^N_j z_j, \sum_{j=1}^{N} \xi^N_j z_j >_X$$

$$ \qquad (1.3.12)$$

$$= \sum_{j,k=1}^{N} \xi^N_j \overline{\xi^N_k} <z_j,z_k>_X = \sum_{j,k=1}^{N} c_j \sigma^N_{j,k} \overline{c_k}$$

with $\sigma^N_{j,k}$ being defined by (1.2.6).

From (1.3.10) we infer that, for all $N \epsilon \mathbb{N}$,

$$\sum_{j,k=1}^{N} a_j <z_j , z_k>_X \bar{a}_k \leq \mu \sum_{j=1}^{N} |a_j|^2$$

for all $(a_1, \ldots, a_N) \epsilon \mathbb{R}^N$ or $\mathbb{C}^N$ which implies

$$\sum_{j=1}^{N} b_j \sigma_{j,k}^{N} \bar{b}_k \geq \frac{1}{\mu} \sum_{j=1}^{N} |b_j|^2$$

for all $(b_1, \ldots, b_N) \epsilon \mathbb{R}^N$ or $\mathbb{C}^N$.

So (1.2.9) and (1.3.12) imply

$$\sum_{j=1}^{N} |c_j|^2 \leq \mu \ |u|^2$$

for all $N \epsilon \mathbb{N}$ and in turn (1.3.11) which completes the proof.

As a consequence of Lemma 1.3.7 we immediately see that by

$$S(x) = \sum_{j=1}^{\infty} <x,z_j>y_j , \quad x \epsilon X, \tag{1.3.13}$$

a linear mapping from X into Y is defined. This mapping is called a <u>nuclear operator</u> from X into Y.

In order to determine the adjoint operator S^* of S we take any $y \epsilon Y = Y^*$. Then, for every $x \epsilon X = D(S)$, it follows that

$$<y,S(x)>_Y = \sum_{j=1}^{\infty} \overline{<x,z_j>}_X <y,y_j>_Y = < \sum_{j=1}^{\infty} <y,y_j>_Y z_j , x>_X$$

because of (1.3.10) and $(<y,y_j>_Y)_{j \epsilon \mathbb{N}} \epsilon l^2$.

Therefore $D(S^*) = Y^* = Y$ and

$$S^*(y) = \sum_{j=1}^{\infty} <y,y_j>_Y z_j \epsilon X \tag{1.3.14}$$

for all $y \epsilon Y$. Since $D(S^*) = Y^* = Y$ is total, it follows from the Corollary of Lemma 1.3.2 that S (1.3.13) is continuous. By Theorem 1.3.5 we obtain

<u>Theorem 1.3.8</u>: The operator S (1.3.13) maps X onto Y, if and only if there exists a constant $\lambda > 0$ such that

$$|y|_Y \leq \lambda \ | \ \sum_{j=1}^{\infty} <y,y_j>_Y \ z_j|_X \quad \text{for all } y \in Y. \tag{1.3.15}$$

As we shall see in Section 1.4 the following situation occurs in problems of null-controllability: Let Y_1 be a linear subspace of Y of the form

$$Y_1 = \{y \in Y| \ \sum_{j=1}^{\infty} \alpha_j \ |<y,y_j>_Y|^2 < \infty\}$$

where $(\alpha_j)_{j \in N}$ is a sequence of reals with

$$\alpha_j \geq \alpha > 0 \quad \text{for all } j \in \mathbb{N} \text{ and } \lim_{j \to \infty} \alpha_j = \infty.$$

If we introduce a norm in Y_1 by defining

$$|y|_{Y_1} = (\sum_{j=1}^{\infty} \alpha_j \ |<y,y_j>_Y|^2)^{1/2}, \quad y \in Y_1,$$

then we have

$$|y|_Y \leq \frac{1}{\alpha} |y|_{Y_1} \quad \text{for all } y \in Y.$$

The dual space Y_1^* of Y_1 provided with $|.|_{Y_1}$ as norm consists of all linear functionals $y^*: Y_1 \to \mathbb{R}$ or $\mathbb{C}$ which are of the form

$$y^*(y) = \sum_{j=1}^{\infty} \overline{<y,y_j>}_Y \ y^*(y_j), \quad y \in Y_1,$$

and satisfy

$$\sum_{j=1}^{\infty} \frac{1}{\alpha_j} |y^*(y_j)|^2 < \infty.$$

Let us assume that S (1.3.13) maps X into Y_1 and that

$$\sum_{j=1}^{\infty} y^*(y_j)z_j \in X \quad \text{for all } y^* \in Y_1^*.$$

Then the adjoint operator S^* of $S : X \to Y_1$ is given by

$$S^*(y^*) = \sum_{j=1}^{\infty} y^*(y_j)z_j, \quad y^* \in Y_1^*.$$

By Lemma 1.3.1 the mapping $S : X \to Y_1$ is closed and by Theorems 1.3.3 and 1.3.4 S maps X onto Y_1, if and only if there is a constant $\lambda > 0$ such that

$$\|y^*\|_{Y_1^*} \leq \lambda \ | \ \sum_{j=1}^{\infty} y^*(y_j) z_j \|_X \quad \text{for all } y^* \epsilon Y_1^*$$

where

$$\|y^*\|_{Y_1^*} = (\sum_{j=1}^{\infty} \frac{1}{\alpha_1} | y^*(y_j) |^2)^{1/2}.$$

Before we apply Theorem 1.3.6 concerning approximate solvability of (1.3.1) for all $y \epsilon Y$ we prove

Lemma 1.3.9: Let the sequence $(z_j)_{j \epsilon \, IN}$ be minimal in X (see Section 1.2.1 for the Definition). Then the adjoint operator S^* (1.3.14) of S (1.3.13) is injective.

Proof: Assume $S^*(y) = \Theta_X$ = zero element of X. Then

$$\langle y, y_j \rangle_Y = 0 \quad \text{for all } y \epsilon Y$$

because $\langle y, y_k \rangle \neq 0$ for some $k \epsilon$ N would imply

$$z_k = \sum_{\substack{j=1 \\ j \neq k}}^{\infty} - \frac{\langle y, y_j \rangle_Y}{\langle y, y_k \rangle_Y} z_j$$

contradicting the minimality of $(z_j)_{j \epsilon \, IN}$. Now $y = \Theta_Y$ = zero element of Y follows from the completeness of $(y_j)_{j \epsilon \, IN}$ in Y.

By Theorem 1.3.6 we have

Theorem 1.3.10: If $(z_j)_{j \epsilon \, IN}$ is minimal, then the equation (1.3.1) with S being defined by (1.3.13) is approximately solvable for every $y \epsilon Y$, i.e., for every $y \epsilon Y$ there is a sequence (x_n) in X such that (1.3.8) holds.

This result can also be proved directly by applying Theorem 1.2.5 which guarantees the existence of a biorthonormal sequence $(v_k)_{k \epsilon \, IN}$ of $(z_j)_{j \epsilon \, IN}$ in X with

$$\langle v_k, z_j \rangle = \delta_{kj} = \text{Kronecker's symbol for all } j, k \in \mathbb{N}.$$

Let $y \in Y$ be given. Then we define

$$x_n = \sum_{j=1}^{n} \langle y, y_j \rangle v_j \quad \text{for every } n \in \mathbb{N}$$

and conclude

$$S(x_n) = \sum_{j=1}^{n} \langle y, z_j \rangle y_j$$

which implies (1.3.8).

The solvability of (1.3.1) for S defined by (1.3.13) is equivalent to the solvability of a moment problem which is the content of

Lemma 1.3.11: The equation (1.3.1) with S being defined by (1.3.13) has a solution for every $y \in Y$, if and only if the system

$$\langle x, z_j \rangle_X = c_j, \quad j \in \mathbb{N}, \tag{1.3.16}$$

of moment equations has a solution for every sequence $c = (c_j)_{j \in \mathbb{N}} \in l^2$.

The statement of this lemma simply follows from the fact that, due to the completeness of Y, by $y \to (\langle y, y_j \rangle_Y)_{j \in \mathbb{N}}$, $y \in Y$, an isometric isomorphism from Y onto l^2 is defined.

As a consequence of Lemma 1.3.11 we obtain by Theorem 1.3.8 a necessary and sufficient condition for the solvability of the moment problem (1.3.16) for every sequence $(c_j)_{j \in \mathbb{N}} \in l^2$. The sufficiency of (1.3.15) also follows from Theorem 1.2.3, if X is separable. The necessicity of (1.3.15) could not be proved in Section 1.2.1 because the condition (1.3.10) on the sequence $(z_j)_{j \in \mathbb{N}}$ was not required there. We shall see in Section 1.4.1, however, that for control problems (1.3.10) is often satisfied in cases where the solvability of (1.3.16) can be guaranteed. In fact we shall see that part of the results in Section 1.2.3 can also be obtained from the results in this section.

1.4. Application to One-Dimensional Vibrations.

1.4.1. Distributed Control.

We pick up the problem of null-controllability by distributed controls als introduced in Section 1.1.1. For a given time $T>0$ and $u \in L^2(0,T)$ we define

$$
S(u) = \begin{pmatrix} \sum\limits_{j=1}^{\infty} \dfrac{h_j}{\sqrt{\lambda_j}} \int\limits_0^T u(t) \sin \sqrt{\lambda_j}(T-t) \, dt \, e_j \\[2em] \sum\limits_{j=1}^{\infty} h_j \int\limits_0^T u(t) \cos \sqrt{\lambda_j}(T-t) \, dt \, e_j \end{pmatrix} \tag{1.4.1}
$$

where $(\lambda_j)_{j \in \mathbb{N}}$ is the sequence of eigenvalues and $(e_j)_{j \in \mathbb{N}}$ a corresponding complete orthonormal sequence of eigenfunctions of the self-adjoint and positive definite differential operator L in (1.1.1). The sequence $(h_j)_{j \in \mathbb{N}}$ is defined by (1.1.16) with $r \in L^2(0,1)$ on the right-hand side of (1.1.1).

If we put

$$
y = \begin{pmatrix} -\sum\limits_{j=1}^{\infty} (a_j \cos \sqrt{\lambda_j}T + \dfrac{b_j}{\sqrt{\lambda_j}} \sin \sqrt{\lambda_j}T) e_j \\[2em] -\sum\limits_{j=1}^{\infty} \sqrt{\lambda_j}(-a_j \sin \sqrt{\lambda_j}T + \dfrac{b_j}{\sqrt{\lambda_j}} \cos \sqrt{\lambda_j}T) e_j \end{pmatrix} \tag{1.4.2}
$$

with $(a_j)_{j \in \mathbb{N}}$ and $(b_j)_{j \in \mathbb{N}}$ being defined by (1.1.16), then (1.1.5) is equivalent to $S(u) = y$.

So the question whether null-controllability is possible within the time-interval $[0,T]$ for any choice of an initial state $(y_0, y_1) \in E \times L^2(0,1)$ and E being defined by (1.1.9) amounts to the question of solvability of $S(u) = y$ for every y of the form (1.4.2).

From $(y_0, y_1) \in E \times L^2(0,1)$ it follows that y (1.4.2) belongs to the Hilbert space $Y = E \times L^2(0,1)$ equipped with the norm

$$
|(y_1, y_2)|_Y = (|y_1|_E^2 + |y_2|_{L^2(0,1)}^2)^{1/2} ,
$$

$y_1 \in E$, $y_2 \in L^2(0,1)$ with $|.|_E$ defined by (1.1.10).

Let $X = L^2(0,T)$. Then by (1.4.1) a continuous linear mapping S from X into Y is defined and null-controllability for every choice of $(y_0, y_1) \in E \times L^2(0,1)$ holds true, if S maps X onto Y. Let us define, for every $j \in \mathbb{N}$ and $t \in [0,T]$,

$$z_{2j-1}(t) = h_j \sin \sqrt{\lambda_j} t, \quad z_{2j}(t) = h_j \cos \sqrt{\lambda_j} t,$$

$$y_{2j-1} = \begin{pmatrix} \frac{1}{\sqrt{\lambda_j}} e_j \\ 0 \end{pmatrix}, \quad y_{2j} = \begin{pmatrix} 0 \\ e_j \end{pmatrix}, \qquad (1.4.3)$$

and $x(t) = u(T-t)$, $u \in L^2(0,T)$. Then $x \in L^2(0,T)$ and

$$S(u) = S(x) = \sum_{j=1}^{\infty} \langle x, z_j \rangle_X y_j \qquad (1.4.1')$$

is a nuclear operator (see (1.3.13)). Because of

$$|z_j|_X^2 \leq h_j^2 T \text{ for all } j \in \mathbb{N}$$

the condition (1.3.10) is satisfied. Also $(y_j)_{j \in \mathbb{N}}$ is a complete orthonormal sequence in $E \times L^2(0,1)$ so that the basic assumptions of Section 1.3.3 are satisfied here.

The question now arises under which conditions Theorem 1.3.8 can be applied in order to prove that S (1.4.1') maps $X = L^2(0,T)$ onto $Y = E \times L^2(0,1)$. This question, however, is not senseful because S cannot map X onto Y. In view of condition (1.1.27) which plays a fundamental role in Sections 1.1.2, 1.1.4 and 1.2.3 this is not so amazing. Therefore it is reasonable to replace Y by a suitable subspace Y_1 such that every y defined by (1.4.2) is in Y_1, if $(y_0, y_1) \in Y_1$, and that by suitable conditions on the sequence $(z_j)_{j \in \mathbb{N}}$ in (1.4.3) it can be ensured that S maps X onto Y_1. In view of condition (1.1.27) we choose $Y_1 = E_r \times F_r$ where

$$E_r = \{ y \in E \mid \sum_{j=1}^{\infty} \frac{1}{h_j^2} \langle y, e_j \rangle_E^2 < \infty \} \qquad (1.4.4)$$

and

$$F_r = \{ y \in L^2(0,1) \mid \sum_{j=1}^{\infty} \frac{1}{h_j^2} \langle y, e_j \rangle_{L^2(0,1)}^2 < \infty \} \qquad (1.4.5)$$

equipped with the norms

$$|y|_{E_r} = \left(\sum_{j=1}^{\infty} \frac{1}{h_j^2} <y,e_j>_E^2 \right)^{1/2}, \quad y \in E_r \tag{1.4.6}$$

and

$$|y|_{F_r} = \left(\sum_{j=1}^{\infty} \frac{1}{h_j^2} <y,e_j>_{L^2(0,1)}^2 \right)^{1/2}, \quad y \in F_r, \tag{1.4.7}$$

respectively.

If we define

$$\alpha_j = \frac{1}{h_j^2} \quad \text{for all } j \in \mathbb{N},$$

then, with $(y_j)_{j \in \mathbb{N}}$ according to (1.4.3),

$$Y_1 = \{y \in Y | \sum_{j=1}^{\infty} \alpha_j |<y,y_j>_Y|^2 < \infty\}$$

and we are in the situation following Theorem 1.3.8 because $(h_j^2)_{j \in \mathbb{N}}$ is a null-sequence, hence bounded by some $\gamma > 0$ from above, which implies

$$\alpha_j \geq \alpha = \frac{1}{\gamma} > 0 \quad \text{for all } j \in \mathbb{N}.$$

In order to apply the results in Section 1.3.3 we have to make sure that S (1.4.1') maps X into Y_1 and that

$$\sum_{j=1}^{\infty} y^*(y_j) z_j \in X \quad \text{for all } y^* \in Y_1^*. \tag{1.4.8}$$

Lemma 1.4.1: If there is a constant $\gamma > 0$ such that

$$\int_0^T |\sum_{j=1}^{\infty} a_{2j-1} \sin \sqrt{\bar{\lambda}_j} t + a_{2j} \cos \sqrt{\bar{\lambda}_j} t|^2 \, dt \leq \gamma \sum_{j=1}^{\infty} |a_j|^2 \tag{1.4.9}$$

for all sequences $(a_j)_{j \in \mathbb{N}} \in l^2$,

then S (1.4.1') maps X into Y_1 and (1.4.8) is satisfied.

Proof: $S(X) \subseteq Y_1$ is a consequence of

$$\sum_{j=1}^{\infty} \alpha_j <S(x),y_j>_Y^2 = \sum_{j=1}^{\infty} (\int_0^T x(t) \sin \sqrt{\bar{\lambda}_j}t \, dt)^2$$

$$+ (\int_0^T x(t) \cos \sqrt{\bar{\lambda}_j}t \, dt)^2$$

and Lemma 1.3.7. The condition (1.4.8) follows from (1.4.9) in connection with

$$\sum_{j=1}^{\infty} y^*(y_j)z_j(t) = \sum_{j=1}^{\infty} (y^*(y_{2j-1})h_j \sin \sqrt{\bar{\lambda}_j}t$$

$$+ y^*(y_{2j})h_j \cos \sqrt{\bar{\lambda}_j}t)$$

and

$$\sum_{j=1}^{\infty} |y^*(y_j)|^2 h_j^2 = \sum_{j=1}^{\infty} \frac{1}{\alpha_j} |y^*(y_j)|^2 < \infty$$

for all $y^* \epsilon Y_1^*$.

Lemma 1.4.2: If there is a constant $\lambda > 0$ such that

$$\sum_{j=1}^{\infty} |a_j|^2 \leq \lambda^2 \int_0^T |\sum_{j=1}^{\infty} a_{2j-1} \sin \sqrt{\bar{\lambda}_j}t + a_{2j} \cos \sqrt{\bar{\lambda}_j}t|^2 \, dt$$

$$(1.4.10)$$

for all sequences $(a_j)_{j \epsilon \text{IN}} \epsilon l^2,$

then

$$|y^*|_{Y_1^*} \leq \lambda \ | \sum_{j=1}^{\infty} y^*(y_j)z_j|_X \quad \text{for all } y^* \epsilon Y_1^* \qquad (1.4.11)$$

which is equivalent to $S(X) = Y_1$.

Proof: (1.4.11) is an immediate consequence of (1.4.10) and the definition (1.4.3) of the sequence $(z_j)_{j \epsilon \text{IN}}$ in connection with

$$|y^*|_{Y_1^*} = (\sum_{j=1}^{\infty} \frac{1}{\alpha_j} |y^*(y_j)|^2)^{1/2}$$

$$= (\sum_{j=1}^{\infty} h_j^2 |y^*(y_j)|^2)^{1/2} \quad \text{for } y^* \epsilon Y_1^*.$$

Before we formulate a condition which guarantees (1.4.9) for some $\gamma > 0$ and (1.4.10) for some $\lambda > 0$ we prove

<u>Lemma 1.4.3:</u> Let $a_{-N}, \ldots, a_{-1}, a_0, a_1, \ldots, a_N$ arbitrary complex numbers and let $\omega_{-N}, \ldots, \omega_{-1}, \omega_0, \omega_1, \ldots, \omega_N$ be reals such that

$$\omega_j - \omega_{j-1} \geq \lambda \quad \text{for} - N < j \leq N \tag{1.4.12}$$

and some $\lambda > 0$. Then for each $\varepsilon > 0$ and $T = \dfrac{\pi + \varepsilon}{\lambda}$ it follows that

$$\int_{-\frac{T}{2}}^{\frac{T}{2}} \left| \sum_{j=-N}^{N} a_j e^{-i\omega_j t} \right|^2 dt \leq \frac{6\pi\sqrt{2}}{T} \sum_{j=-N}^{N} |a_j|^2 \tag{1.4.13}$$

<u>Proof:</u> We proceed as in the proof of Theorem 1.2.9 and define $k = k(t)$, $t \in \mathbb{R}$, and $K = K(\omega)$, $\omega \in \mathbb{R}$, as there. For

$$f(t) = \sum_{j=-N}^{N} a_j e^{-i\omega t}, \quad t \in \mathbb{R},$$

we then have

$$\int_{-\infty}^{\infty} k(t) |f(t)|^2 dt \leq \sum_{j=-N}^{N} \sum_{k=-N}^{N} \frac{|a_j|^2 + |a_k|^2}{2} |K(\omega_j - \omega_k)|$$

$$= K(0) \sum_{j=-N}^{N} |a_j|^2 + \sum_{j=-N}^{N} |a_j|^2 \sum_{\substack{k=-N \\ k \neq j}}^{N} |K(\omega_j - \omega_k)|$$

As seen in the proof of Theorem 1.2.9 we may assume $T = \pi$ and $\lambda = \dfrac{\pi + \varepsilon}{\pi}$ which implies

$$\sum_{\substack{k=-N \\ k \neq j}}^{N} |K(\omega_j - \omega_k)| < \frac{2}{\lambda^2}.$$

Since $K(0) = 4$, it follows that

$$\int_{-\infty}^{\infty} k(t) |f(t)|^2 dt \leq \left(4 + \frac{2}{\lambda^2}\right) \sum_{j=-N}^{N} |a_j|^2 < 6 \sum_{j=-N}^{N} |a_j|^2.$$

Now

$$\frac{1}{2}\sqrt{2} \int_{-\frac{\pi}{2}}^{\frac{\pi}{2}} |f(t)|^2 dt \leq \int_{-\frac{\pi}{2}}^{\frac{\pi}{2}} k(t) |f(t)|^2 dt \leq 6 \sum_{j=-N}^{N} |a_j|^2$$

from which (1.4.13) for $T = \pi$ follows which completes the proof.

As a consequence of Lemma 1.4.3 we obtain

Lemma 1.4.4: Let $\lambda_0 = 0$ and assume that, for a given $T>0$, there is some $\varepsilon>0$ such that

$$\sqrt{\lambda_j} - \sqrt{\lambda_{j-1}} \geq \frac{\pi + \varepsilon}{T} \quad \text{for all } j \in \mathbb{N}. \tag{1.4.14}$$

Then, for every real sequence $(a_j)_{j \in \mathbb{N}}$ in l^2 it follows that

$$\int_0^T \left| \sum_{j=1}^\infty a_{2j-1}\sin \sqrt{\lambda_j} t + a_{2j}\cos \sqrt{\lambda_j} t \right|^2 dt \leq \frac{3\pi\sqrt{2}}{T} \sum_{j=0}^\infty |a_j|^2 \tag{1.4.9'}$$

Proof: We define $\omega_0 = 0$, $\omega_j = \sqrt{\lambda_j}$, $\omega_{-j} = - \omega_j$,

$$c_0 = 0, \quad c_j = \frac{1}{2}(a_{2j} + ia_{2j-1})e^{-i\omega_j\frac{T}{2}},$$

and $c_{-j} = \bar{c}_j$ for all $j \in \mathbb{N}$.

Then, for every $N \in \mathbb{N}$, it follows that

$$\sum_{j=-N}^N |c_j|^2 = \frac{1}{2} \sum_{j=0}^{2N} |a_j|^2$$

and

$$\sum_{j=-N}^N c_j e^{-i\omega_j(t-\frac{T}{2})} = \sum_{j=1}^N a_{2j-1} \sin \sqrt{\lambda_j} t + a_{2j} \cos \sqrt{\lambda_j} t,$$

consequently, by Lemma 1.4.3,

$$\int_0^T \left| \sum_{j=1}^N a_{2j-1} \sin \sqrt{\lambda_j} t + a_{2j} \cos \sqrt{\lambda_j} t \right|^2 dt$$

$$= \int_0^T \left| \sum_{j=-N}^N c_j e^{i\omega_j(t-\frac{T}{2})} \right|^2 dt = \int_{-\frac{T}{2}}^{\frac{T}{2}} \left| \sum_{j=-N}^N c_j e^{-i\omega_j t} \right|^2 dt$$

$$\leq \frac{6\pi\sqrt{2}}{T} \sum_{j=-N}^N |c_j|^2 = \frac{3\pi\sqrt{2}}{T} \sum_{j=-N}^N |a_j|^2$$

From this (1.4.9') immediately follows.

In a similar way Theorem 1.2.9 implies

Lemma 1.4.5: Let $\lambda_0 = 0$ and assume that, for a given T>0, there is some $\varepsilon>0$ such that

$$\sqrt{\lambda_j} - \sqrt{\lambda_{j-1}} \geq \frac{2\pi + \varepsilon}{T} \qquad \text{for all } j \in \mathbb{N}. \qquad (1.4.15)$$

Then, for every real sequence $(a_j)_{j \in \mathbb{N}}$ in l^2 it follows that

$$\sum_{j=1}^{\infty} |a_j|^2 \leq \frac{A(\varepsilon)}{T} \int_0^T |\sum_{j=1}^{\infty} a_{2j-1} \sin \sqrt{\lambda_j}t + a_{2j} \cos \sqrt{\lambda_j}t|^2 \, dt \qquad (1.4.10')$$

with $A(\varepsilon)$ defined by (1.2.28).

Summarizing we see, since condition (1.4.14) is implied by (1.4.15), that the latter condition guarantees the assumptions of Lemmas 1.4.4 and 1.4.5 and in turn that S (1.4.1) maps $L^2(0,T)$ onto $E_r \times F_r$ with E_r and F_r being defined by (1.4.4) and (1.4.5), respectively. This result is also contained in Theorem 1.2.19.

1.4.2. Boundary Control.

Here we return to the situation of Section 1.1.3. For a given time T>0 and

$$v \in H_0^2(0,T) = \{v \in H^2(0,T) \mid v(0) = v'(0) = v(T) = v'(T) = 0\}$$

we define

$$\hat{S}(v) = \begin{pmatrix} \sum_{j=1}^{\infty} \frac{h_j}{\sqrt{\lambda_j}} \int_0^T v''(t) \sin \sqrt{\lambda_j}(T-t) \, dt \; e_j \\ \\ \sum_{j=1}^{\infty} h_j \int_0^T v''(t) \cos \sqrt{\lambda_j}(T-t) \, dt \; e_j \end{pmatrix} \qquad (1.4.16)$$

with $(\lambda_j)_{j \in \mathbb{N}}$, $(e_j)_{j \in \mathbb{N}}$, $(h_j)_{j \in \mathbb{N}}$ being defined as in Sections 1.1.1 and 1.1.3. If we again define y by (1.4.2), then (1.1.40) is equivalent with $\hat{S}(v) = y$. If we define T : $H_0^2(0,T) \to L^2(0,T)$ by Tv = v'' a.e. on (0,T), then $\hat{S}(v) = S(Tv)$ with S defined by (1.4.1). By the result at the end of Section 1.4.1 it follows that $\hat{S}$ maps $H_0^2(0,T)$ into $E_r \times F_r$ with E_r defined by (1.4.4) and F_r by (1.4.5), if (1.4.15) holds. In order to show then that $\hat{S}$ is surjective we have to prove, by Theorem 1.3.4, that $\hat{S}$ is closed

and that its adjoint operator $\hat{S}^*$ has a bounded inverse. For
the prove of the closedness of $\hat{S}$ we make use of Lemma 1.3.1.
First we observe that $D(\hat{S}) = H_0^2(0,T)$ is complete with respect
to the norm

$$|v|^2_{H_0^2(0,T)} = <v'',v''>^2_{L^2(0,T)} \, , \quad v\in H_0^2(0,T).\tag{1.4.17}$$

If we introduce the norm

$$|(y_1,y_2)|_{E\times L^2(0,1)} = \left(|y_1|^2_E + |y_2|^2_{L^2(0,1)}\right)^{1/2}, \quad y_1\in E_r, y_2\in F_r,\tag{1.4.18}$$

in $E_r\times F_r$, then

$$|y|_{E\times L^2(0,1)} \leq \gamma \, |y|_{E_r\times F_r} \quad \text{for all } y\in E_r\times F_r$$

where $\gamma>0$ is a constant with

$$h_j^2 \leq \gamma \text{ for all } j\in \mathbb{N}.$$

Since $\hat{S}$ maps $H_0^2(0,T)$ equipped with the norm (1.4.17) continuously
into $E_r\times F_r$ equipped with the norm (1.4.18), it is closed with
respect to the $|\cdot|_{E_r\times F_r}$-norm of $E_r\times F_r$ by Lemma 1.3.1.

In order to compute the adjoint operator $\hat{S}^*$ of $\hat{S}$ we choose
$(y_1^*,y_2^*)\in E_r^*\times F_r^*$ and obtain for $\hat{S}(v) = (\hat{S}_1(v), \hat{S}_2(v))$, $v\in H_0^2(0,T)$

$$y_1^*\hat{S}_1(v) + y_2^*\hat{S}_2(v) = <w,v''>_{L^2(0,T)}$$

where

$$w(\cdot) = \sum_{j=1}^{\infty} \frac{h_j}{\sqrt{\bar\lambda_j}} \, y_1^*(e_j) \sin \sqrt{\bar\lambda_j}\,(T-\cdot) + h_j y_2^*(e_j) \cos \sqrt{\bar\lambda_j}\,(T-\cdot)$$
$$\in L^2(0,T) \tag{1.4.19}$$

if (1.4.15) holds (see Lemma 1.4.4).

Since

$$|<w,v''>_{L^2(0,T)}| \leq |w|_{L^2(0,T)} |v|_{H_0^2(0,T)} \, , \quad v\in H_0^2(0,T),$$

it follows that for every pair $(y_1^*,y_2^*)\in E_r^*\times F_r^*$ by $v\to y_1^*\hat{S}_1(v) +$
$y_2^*\hat{S}_2(v)$ a continuous linear functional is defined on $H_0^2(0,T)$.
Therefore $D(\hat{S}^*) = E_r^*\times F_r^*$ and

$$\hat{S}^*(y_1^*, y_2^*) = \int_0^t (t-s)w(s) \, ds, \quad (y_1^*, y_2^*) \in E_r^* \times F_r^*$$

with w being defined by (1.4.19). $\hat{S}^*$ then maps $E_r^* \times F_r^*$ into $V = \{v \in H^2(0,T) \mid v(0) = v'(0) = 0\}$ which can be identified with the dual space of $H_0^2(0,T)$. Further it follows from Lemma 1.4.5 that, for every pair $(y_1^*, y_2^*) \in E_r^* \times F_r^*$,

$$|\hat{S}^*(y_1^*, y_2^*)|_V^2 = |w|_{L^2(0,T)}^2 \geq \frac{A(\varepsilon)}{T} |(y_1^*, y_2^*)|_{E_r^* \times F_r^*}^2 ,$$

if (1.4.15) is true. By Theorem 1.3.4 then $\hat{S}$ maps $H_0^2(0,T)$ onto $E_r \times F_r$. This result is also contained in Theorem 1.2.21.

1.5. On Time-Minimal Control of Linear Systems.
1.5.1. Existence of Time-Minimal Controls.

The problem of time-minimal null-controllability as being introduced in Sections 1.1.1 and 1.1.3 will be treated here in a general fashion which only makes use of the underlying structure of the problem and continues the considerations of Section 1.3. We therefore consider the following situation: Given two Banach spaces X and Y, let $(S_T)_{T \in [0,\hat{T}]}$ for some $\hat{T} \in (0, \infty]$ be a family of continuous linear mappings from X into Y. Further let y : $[0,\hat{T}] \to Y$ be a given function and for some constant M>0 let

$$U_M = \{u \in X \mid |u|_X \leq M\} \tag{1.5.1}$$

The problem of (restricted) controllability now consists of finding some $T \in [0,\hat{T}]$ and some $u \in U_M$ such that

$$S_T(u) = y(T) \tag{1.5.2}$$

The investigations in Section 1.3 are concerned with the solvability of the equation (1.5.2) for a given fixed $T \in [0,\hat{T}]$ and without the requirement $u \in U_M$. We assume in this section that the problem of controllability has a solution for $T = \hat{T}$, i.e., there exists some $u \in U_M$ such that (1.5.2) holds for $T = \hat{T}$. We are then interested in the problem of time-minimal controllability which consists of finding the smallest time $T = T(M)$

such that there is some $u \epsilon U_M$ which satisfies (1.5.2) for
$T = T(M)$. Every $u \epsilon U_M$ with this property is called a
time-minimal control. The minimum time $T(M)$ is well
defined as

$$T(M) = \inf\{T \epsilon [0,\hat{T}] \mid S_T(u) = y(T) \text{ for some } u \epsilon U_M\}. \qquad (1.5.3)$$

In addition to the above assumptions we require that

1) $S_0(X) = \{\Theta_Y\}$, Θ_Y = zero element of Y.

2) $\displaystyle\lim_{T \to T^*+0} |S_T - S_{T^*}| = 0$ for every $T^* \epsilon [0,\hat{T}]$.

3) The space X is the dual space of a separable Banach space
   Z or a reflexive Banach space itself and, for every
   $T \epsilon [0,\hat{T}]$, the operator S_T maps weak* convergent into
   norm-convergent sequences, i.e., $(u_k)_{k \epsilon X}$ in X, $u \epsilon X$ with
   $u_k \overset{*}{\to} u$ implies that

$$\lim_{k \to \infty} |S_T(u_k) - S_T(u)|_Y = 0.$$

4) $y(0) \neq \Theta_Y$ and

$$\lim_{T \to T^*+0} |y(T) - y(T^*)|_Y = 0 \text{ for every } T^* \epsilon [0,\hat{T}].$$

Then we can prove the following existence statement.

Theorem 1.5.1: There is some $u_M \epsilon U_M$ with

$$S_{T(M)}(u_M) = y(T(M)) \qquad (1.5.4)$$

and $T(M) > 0$.

Proof: By the definition (1.5.3) of $T(M)$ there is a sequence
$(T_k)_{k \epsilon \text{ IN}}$ in $[0,\hat{T}]$ such that $T_k \to T(M) + 0$ and for each k there
exists some $u_k \epsilon U_M$ with

$$S_{T_k}(u_k) = y(T_k)$$

Since U_M is weak* sequentially compact in X, there is a
subsequence (u_{k_i}) and some $u_M \epsilon U_M$ such that $u_{k_i} \overset{*}{\to} u_M$. For each i

we then have

$$|S_{T(M)}(u_M) - y(T(M))|_Y \leq |S_{T(M)}(u_M) - S_{T(M)}(u_{k_i})|_Y$$

$$+ |S_{T(M)}(u_{k_i}) - S_{T_{k_i}}(u_{k_i})|_Y + |y(T_{k_i}) - y(T(M))|_Y$$

$$\leq |S_{T(M)}(u_M) - S_{T(M)}(u_{k_i})|_Y + |S_{T(M)} - S_{T_{k_i}}| \cdot M$$

$$+ |y(T_{k_i}) - y(T(M))|_Y.$$

From $T_{k_i} \to T(M) + 0$ we conclude, by 2) and 4), that

$$\lim_{i \to \infty} |S_{T_{k_i}} - S_{T(M)}| = 0 \text{ and } \lim_{i \to \infty} |y(T_{k_i}) - y(T(M))|_Y = 0$$

and $u_{k_i} \overset{*}{\to} u_M$ implies, by 3), that

$$\lim_{i \to \infty} |S_{T(M)}(u_M) - S_{T(M)}(u_{k_i})|_Y = 0.$$

As a result we obtain (1.5.4).

If we assume $T(M) = 0$, then there is a sequence $(T_k)_{k \in \mathbb{N}}$ in $[0, \hat{T}]$ with $T_k \to 0$ and for each k some $u_k \in U_M$ with $S_{T_k}(u_k) = y(T_k)$ which implies

$$|y(0)|_Y = \lim_{k \to \infty} |y(T_k)|_Y \leq \lim_{k \to \infty} |S_{T_k}| \cdot M = 0,$$

a contradiction to $y(0) \neq 0$ which completes the proof.

1.5.2 A General Maximum-Principle for Minimum Norm Controls.

In this section we consider the equation (1.5.2) for some fixed time $\hat{T} \in (0, \hat{T}]$ and investigate the question under which conditions one can guarantee the existence of minimum norm controls, i.e., of solutions $u \in X$ of (1.5.2) with smallest possible norm $|u|_X$, and how minimum norm controls can be characterized. The reason for this investigation is that, under suitable assumptions, it will be shown in Section 1.5.3 that every time-minimal control is also a minimum norm control for $T = T(M)$ (1.5.3).

Let $W = S_T(X)$ and let $y \in W$ be given. This means that there exists some $u \in X$ with $S_T(u) = y$. We put

$$v_T(y) = \inf\{|u|_X|\ u\epsilon X,\ S_T(u) = y\} \tag{1.5.5}$$

and prove

Theorem 1.5.2: Under the assumption 3) of Theorem 1.5.1 there exists, for every $y\epsilon W$, some $u_{T,y}\epsilon X$ with

$$S_T(u_{T,y}) = y \text{ and } |u_{T,y}| = v_T(y). \tag{1.5.6}$$

Proof: By the definition (1.5.5) of $v_T(y)$ there exists a sequence $(u_k)_{k\epsilon\mathbb{N}}$ with $S_T(u_k) = y$ for all $k\epsilon\mathbb{N}$ and $\lim\limits_{k\to\infty}|u_k|_X = v_T(y)$ which implies that $(u_k)_{k\epsilon\mathbb{N}}$ is bounded in X and therefore has a subsequence $(u_{k_i})_{i\epsilon\mathbb{N}}$ such that $u_{k_i} \overset{*}{\to} u_{T,y}$ for some $u_{T,y}\epsilon X$. Since the function $u\to|u|_X$ is weak* sequentially lower semi-continuous it follows that

$$|u_{T,y}|_X \leq \liminf\limits_{i\to\infty}|u_{k_i}|_X = v_T(y).$$

Moreover, from $u_{k_i} \overset{*}{\to} u_{T,y}$ we infer, by the assumption 3) of Theorem 1.5.1, that

$$S_T(u_{T,y}) = \lim\limits_{i\to\infty}S_T(u_{k_i}) = y \tag{1.5.7}$$

and hence

$$|u_{T,y}|_X = v_T(y)$$

which completes the proof.

Remark: For the proof of Theorem 1.5.2 we only need that S_T maps weak* convergent into weakly convergent sequences, i.e., $(u_k)_{k\epsilon\mathbb{N}}$ in X and $u_k\overset{*}{\to}u$ for some $u\epsilon X$ implies that $S_T(u_k)\to S_T(u)$ which means $\lim\limits_{k\to\infty}y^*(S_T(u_k)) = y^*(S_T(u))$ for all $y^*\epsilon Y^*$.
In this case, instead of (1.5.7), one can conclude that

$$y^*(S_T(u_{T,y})) = \lim\limits_{i\to\infty}y^*(S_T(u_{k_i})) = y^*(y) \text{ for all } y^*\epsilon Y^*$$

which also implies $S_T(u_{T,y}) = y$.

The next Theorem is the basis for a general maximum-principle
for minimum norm controls. Before formulating it we assume
that $W = S_T(X)$ is a subspace of Y which is a Banach space with
respect to a norm $|.|_W$ such that $S_T : X \to W$ is continuous with
respect to the norm $|.|_W$ in W.

Under this assumption we then have

Theorem 1.5.3: For every $y \in W$ there is some $y^* \in W^*$ with $y^* \neq \Theta_W$
such that

$$y^*(y) = \sup\{y^*(S_T(u)) \mid u \in X, \ |u|_X \leq v_T(y)\}. \qquad (1.5.8)$$

Proof: If $y = \Theta_Y$, then (1.5.8) is true for any $y^* \in W^*$ with
$y^* \neq \Theta_{W^*}$. Let $y \neq \Theta_Y$ and put

$$M_y = \{u \in X \mid S_T(u) = y\},$$

$$B_y = \{u \in X \mid |u|_X \leq v_T(y)\}.$$

Then the intersection $M_y \cap \overset{\circ}{B}_y$, $\overset{\circ}{B}_y$ being the interior of B_y, is
empty, by the definition (1.5.5) of $v_T(y)$. A well known
separation theorem for disjoint convex sets guarantees the
existence of some $x^* \in X^*$ with $x^* \neq \Theta_{X^*}$ and some $\alpha \in \mathbb{R}$ such that

$$x^*(x) \leq \alpha \leq x^*(u) \quad \text{for all } x \in B_y \text{ and } u \in M_y.$$

The left inequality implies

$$|x^*| v_T(y) = \sup_{x \in B_y} x^*(x) \leq \alpha$$

and from the right it follows that

$$\alpha \leq \inf_{u \in M_y} x^*(u) \leq |x^*| v_T(y).$$

Hence

$$\alpha = |x^*| v_T(y) = \sup_{x \in B_y} x^*(x) = \inf_{u \in M_y} x^*(u)$$

Take any $\hat{u} \in M_y$. Then $M_y = \hat{u} + V$ where

$$V = \{u \in X \mid S_T(u) = \Theta_Y\}.$$

Furthermore,

$$\alpha - x^*(\hat{u}) \leq x^*(u) \quad \text{for all } u \in V.$$

This implies $x^*(u) = 0$ for all $u \in V$, hence,

$$x^* \in V^\perp = \{v^* \in X^* \mid v^*(v) = 0 \quad \text{for all } v \in V\}, \tag{1.5.9}$$

and $\alpha = x^*(\hat{u})$ for all $\hat{u} \in M_y$.

As a result we obtain

$$x^*(u) = \sup_{x \in B_y} x^*(x) \quad \text{for all } u \in M_y. \tag{1.5.10}$$

Since W is a Banach space, it follows from [4] Satz V.2.2 that $V^\perp = S_T^\times(W^*)$. Therefore (1.5.9) implies the existence of some $y^* \in W^*$ with $y^* \neq \Theta_{W^*}$ such that $x^* = S_T^*(y^*)$ and

$$x^*(u) = y^*(y) \quad \text{for all } u \in M_y.$$

This, in conjunction with (1.5.10), implies (1.5.8) and completes the proof.

As an immediate consequence we have

<u>Theorem 1.5.4</u>: For every $y \in W = S_T(X)$ and every $u_{T,y} \in X$ with (1.5.6) there exists some $y^* \in W^*$ with $y^* \neq \Theta_{W^*}$ which is independent of $u_{T,y}$ and satisfies

$$S_T^*(y^*)(u_{T,y}) = |S_T^*(y^*)| \, |u_{T,y}|_X. \tag{1.5.11}$$

Moreover, $S_T^*(y^*) \neq \Theta_{X^*}$.

<u>Proof:</u> From Theorem 1.5.3 we obtain the existence of some $y^* \in W^*$ with $y^* \neq \Theta_{W^*}$ such that the <u>maximum principle</u>

$$y^*(S_T(u_{T,y})) = \sup\{y^*(S_T(u)) \mid u \in X, \, |u|_X \leq |u_{T,y}|_X\} \tag{1.5.12}$$

holds which is equivalent to (1.5.11) and independent of $u_{T,y}$.

Since $S_T(X) = W$, it follows from Theorem 1.3.3 that $S_T^*(y^*) \neq \Theta_{X^*}$, since $y^* \neq \Theta_{W^*}$.

Conversely we have

Theorem 1.5.5: If for some $\hat{u} \in X$ with $S_T(\hat{u}) = y$ there exists $y^* \in W^*$ with $S_T^*(y^*) \neq \Theta_{X^*}$ and

$$S_T^*(y^*)(\hat{u}) = |S_T^*(y^*)| \ |\hat{u}|_X, \qquad (1.5.11')$$

then it follows that

$$|\hat{u}|_X = v_T(y)$$

Proof: Let $u \in X$ with $S_T(u) = y$ be given. Then

$$|S_T^*(y^*)| \ |\hat{u}|_X = S_T^*(y^*)(\hat{u}) = y^*(S_T(\hat{u})) = y^*(y)$$

$$= y^*(S_T(u)) = S_T^*(y^*)(u) \leq |S_T^*(y^*)| \ |u|_X$$

which implies $|\hat{u}|_X \leq |u|_X$ because of $|S_T^*(y^*)| > 0$.

1.5.3. Reduction of Time-Minimal Controllability to Norm-Minimal Controllability.

We return to the situation of Section 1.5.1. In addition to the assumptions 1) - 4) of Theorem 1.5.1 we require the following condition:

5) There exists some $T_0 \in [0,\hat{T})$ and a subspace W of Y such that

$$S_T(X) = W \text{ and } y(T) \in W \quad \text{for all } T \in (T_0, \hat{T}].$$

As a consequence of this assumption the set

$$M(T) = \{u \in X| \ S_T(u) = y(T)\} \qquad (1.5.13)$$

is non-empty and

$$v_T = \inf\{|u|_X| \ u \in M(T)\} \qquad (1.5.14)$$

is well-defined for every $T \in (T_0, \hat{T}]$.

Moreover, Theorem 1.5.2 guarantees, for each $T \in (T_0, \hat{T}]$, the existence of some $u_T \in M(T)$ with $|u_T|_X = v_T$. There is a natural way of introducing a norm in W such that W becomes a Banach space with this norm and every S_T, $T \in (T_0, \hat{T}]$, is continuous with respect to this norm. For this purpose we define, for every

$T \epsilon (T_0, \hat{T}]$, a norm $|.|_T$ in W by putting

$$|y|_T = v_T(y) \quad \text{for all } y \epsilon W \tag{1.5.15}$$

where $v_T(y)$ is defined by (1.5.5). It can be shown that $|.|_T$ has all the properties of a norm and that W is complete with respect to this norm. Next we assume that

$$T_0 < T_1 \le T_2 \le \hat{T} \implies |y|_{T_2} \le |y|_{T_1} \quad \text{for all } y \epsilon W. \tag{1.5.16}$$

Then all the norms (1.5.15) are equivalent which can be seen as follows: Let i be the identical mapping from W equipped with $|.|_{T_2}$ onto W equipped with $|.|_{T_1}$ where T_1 and T_2 are chosen as in (1.5.16). The graph $\{(y, iy)| y \epsilon W\}$ of i in W×W equipped with $\max\{|.|_{T_1}, |.|_{T_2}\}$ as norm is closed and therefore i is continuous by the closed graph theorem. This implies the existence of a constant $\lambda(T_1, T_2) > 0$ such that

$$|y|_{T_1} \le \lambda(T_1, T_2)|y|_{T_2} \quad \text{for all } y \epsilon W$$

which together with (1.5.16) ensures the equivalence of $|.|_{T_1}$ and $|.|_{T_2}$.

On putting $\lambda(T) = \lambda(T, \hat{T})$ for all $T \epsilon (T_0, \hat{T}]$ we have in particular that

$$|y|_T \le \lambda(T)|y|_{\hat{T}} \quad \text{for all } y \epsilon W. \tag{1.5.17}$$

If one introduces a norm in W, for instance, by

$$|y|_W = |y|_{\hat{T}} \quad \text{for all } y \epsilon W$$

with $|.|_{\hat{T}}$ being defined by (1.5.15), then W becomes a Banach space with respect to this norm and every S_T, $T \epsilon (T_0, \hat{T}]$, maps X continuously onto W. This is a consequence of

$$|S_T(u)|_W \le |S_T(u)|_T = \inf\{|\tilde{u}|_X| S_T(\tilde{u}) = S_T(u)\} \le |u|_X$$

for all $u \epsilon X$.

In addition to 5) and (1.5.16) we assume

6) For every $T^* \epsilon (T_0, \hat{T}]$ it is true that

$$\lim_{T \to T^*-0} |y(T) - y(T^*)|_{\hat{T}} = 0 \qquad (1.5.18)$$

and

$$\lim_{T \to T^*-0} |y|_T = |y|_{T^*} \quad \text{for every } y \epsilon W. \qquad (1.5.19)$$

After these preparations we assume that the problem of controllability has a solution for $T = \hat{T}$. If the assumptions 1) - 4) of Theorem 1.5.1 are satisfied, then it follows immediately that $v_{T(M)} \leq M$ where $T(M)$ is the minimal time defined by (1.5.3) and $v_{T(M)}$ is defined by (1.5.5) for $T = T(M)$.

Conversely we can prove

Theorem 1.5.6: Under the above assumptions 5), 6) and (1.5.16) it follows that $v_{T(M)} \geq M$, if $T(M) > T_0$.

Proof: We assume that $v_{T(M)} < M$ and put $\delta = \dfrac{M - v_{T(M)}}{2}$. From (1.5.16) and (1.5.19) it follows that

$$|y(T(M))|_{T(M)} \leq |y(T(M))|_{T(M)-\epsilon} < |y(T(M))|_{T(M)} + \delta$$

$$= v_{T(M)} + \delta = M - 2\delta + \delta = M - \delta$$

for $\epsilon \epsilon (0, T(M) - T_0)$ sufficiently small. Therefore there exists some $u_1 \epsilon X$ such that

$$y(T(M)) = S_{T(M)-\epsilon}(u_1) \text{ and } |u_1|_X \leq M - \delta.$$

For $\epsilon \epsilon (0, \tilde{T}_0)$, $\tilde{T}_0 = \frac{1}{2}(T(M) - T_0)$, sufficiently small we further conclude by (1.5.16), (1.5.17), (1.5.18) that

$$|y(T(M) - \epsilon) - y(T(M))|_{T(M)-\epsilon} \leq |y(T(M) - \epsilon) - y(T(M))|_{\tilde{T}_0}$$

$$\leq \lambda(T_{\tilde{\delta}}) |y(T(M) - \epsilon) - y(T(M))|_{\hat{T}} < \delta.$$

Let $u_2 \epsilon X$ be such that $S_{T(M)-\epsilon}(u_2) = y(T(M) - \epsilon) - y(T(M))$ and $|u_2|_X \leq \delta$. Then $|u_1 + u_2|_X \leq M$ and $S_{T(M)-\epsilon}(u_1 + u_2) = y(T(M) - \epsilon)$ which contradicts the definition (1.5.3) of $T(M)$. Hence the assumption $v_{T(M)} < M$ is false which completes the proof.

we obtain

$$|\tilde{S}_T(u) - \tilde{S}_{T^*}(u)|_Y^2 \le (\sum_{j=1}^{\infty} h_j^2)(T-T^*)|u|_X^2$$

which implies

$$|\tilde{S}_T - \tilde{S}_{T^*}| \le (\sum_{j=1}^{\infty} h_j^2)^{1/2}\sqrt{T-T^*}$$

and thus guarantees 2).

As to assumption 3) we first observe that X is a separable Hilbert space and thereby its own dual space. Weak* convergence in X is equivalent to weak convergence. In order to verify 3) it suffices to show that S_T (1.4.1) is compact for every $T\in[0,\hat{T}]$. For this purpose we define, for every $N\in\mathbb{N}$,

$$\tilde{S}_T^N(u) = \begin{pmatrix} \sum_{j=1}^{N} \dfrac{h_j}{\sqrt{\tilde{\lambda}_j}} \int_0^T u(t)\sin\sqrt{\tilde{\lambda}_j}\, t\, dt\, e_j \\[2mm] \sum_{j=1}^{N} h_j \int_0^T u(t)\cos\sqrt{\tilde{\lambda}_j}\, t\, dt\, e_j \end{pmatrix} \qquad (1.4.1)_N$$

$\in X$. Since $\tilde{S}_T^N$ has a finite-dimensional range in Y, it is compact for every N. Now, for every $u\in X$, we have

$$|\tilde{S}_T(u) - \tilde{S}_T^N(u)|_Y^2 \le (\sum_{j=N+1}^{\infty} h_j^2)T|u|_X^2$$

which implies

$$|\tilde{S}_T - \tilde{S}_T^N| \le (\sum_{j=N+1}^{\infty} h_j^2)^{1/2}T^{1/2} \to 0 \quad \text{as } N\to\infty.$$

This together with the compactness of every $\tilde{S}_T^N$ ensures the compactness of $\tilde{S}_T$ and thus the assumption 3).

Before we can apply Theorem 1.5.1 we have to show that the minimum times T(M) defined by (1.1.7) and (1.5.3) coincide. In view of the above considerations it suffices to prove that the following two sets A and B defined by

$$A = \{T\in[0,\hat{T}]|\ \tilde{S}_T(u) = y \text{ for some } u\in U_M\ (1.5.1)\}$$

and

$$B = \{T\in[0,\hat{T}]|\ \tilde{S}_T(u) = y \text{ for some } u\in L^2(0,T)$$
$$\text{with } |u|_{L^2(0,T)} \le M\},$$

As a consequence of Theorems 1.5.4 and 1.5.6 we obtain

Theorem 1.5.7: Let the assumptions 1) - 4) of Theorem 1.5.1 hold together with the above assumptions 5), 6) and (1.5.16). If $T(M) > T_0$, then

a) $v_{T(M)} = M.$ (1.5.20)

b) Each control $u_M\in U_M$ with (1.5.4) satisfies

$$S^*_{T(M)}(y^*)(u_M) = |S^*_{T(M)}(y^*)|.M \qquad (1.5.21)$$

for some $y^*\in W^*$ with $S^*_{T(M)}(y^*) \neq \Theta_{X^*}$ which is independent of u_M where W^* is the dual space of $W = S_{T(M)}(X)$ being equipped with the norm $|y|_W = |y|_\wedge$, $y\in W$, and $S^*_{T(M)}$ is the adjoint operator of $S_{T(M)}$ considered as a continuous linear mapping from X onto W with respect to the norm $|.|_W$.

Remark: The statement (1.5.21) is equivalent with the maximum-principle

$$S^*_{T(M)}(y^*)(u_M) = \sup\{S^*_{T(M)}(y^*)(u)|\ u\in U_M\}. \qquad (1.5.22)$$

For the remainder of this section we assume the function $y:[0,\hat{T}]\to Y$ to be constant, in particular,

$$y(T) = \hat{y}\in W, \quad \hat{y} \neq \Theta_W, \quad \text{for all } T\in[0,\hat{T}]. \qquad (1.5.23)$$

Then the first part of the above assumption 6) is trivially satisfied. The second part is ensured by

Lemma 1.5.8: For each pair T_1, T_2 with $0<T_1\le T_2\le\hat{T}$ and each $u\in X$ we assume the existence of two elements $u^1_{T_1}$, $u^2_{T_1}\in X$ such that

$$S_{T_2}(u) = S_{T_2}(u^1_{T_1}) + S_{T_2}(u^2_{T_1}), \quad |u^1_{T_1}|_X \le |u|_X,$$

and (1.5.24)

$$\lim_{T_1\to T_2-0} |S_{T_2}(u^2_{T_1})|_{T_2} = 0.$$

Then (1.5.19) is satisfied for every $T^*\in(T_0,\hat{T})$.

Proof: Let $\varepsilon_0 > 0$ be given such that $T_0 < T^* - \varepsilon_0$. Then, for every $\varepsilon \in (0, \varepsilon_0]$ and every $y \in W$, we have

$$|y|_{T^*} \leq |y|_{T^* - \varepsilon} \leq |y|_{T^* - \varepsilon_0} \leq \lambda(T^* - \varepsilon_0, T^*)|y|_{T^*} \quad \text{for some}$$

$\lambda(T^* - \varepsilon_0, T^*) > 0$ which is independent of y. Let us choose $u \in X$ with $S_{T^*}(u) = y$ and $|u|_X = |y|_{T^*}$. Then, for $T_2 = T^*$ and $T_1 = T^* - \varepsilon$, it follows from (1.5.24) that

$$|y|_{T^* - \varepsilon} \leq |S_{T^*}(u^1)|_{T^* - \varepsilon} + |S_{T^*}(u^2)|_{T^* - \varepsilon}$$

$$\leq |u^1|_{T^* - \varepsilon \, X} + \lambda(T^* - \varepsilon_0, T^*)|S_{T^*}(u^2)|_{T^* - \varepsilon \, T^*}$$

$$\leq |u|_X + \lambda(T^* - \varepsilon_0, T^*)|S_{T^*}(u^2)|_{T^* - \varepsilon \, T^*}$$

$$= |y|_{T^*} + \lambda(T^* - \varepsilon_0, T^*)|S_{T^*}(u^2)|_{T^* - \varepsilon \, T^*}$$

Therefore

$$0 \leq |y|_{T^* - \varepsilon} - |y|_{T^*} \leq \lambda(T^* - \varepsilon_0 T^*)|S_{T^*}(u^2)|_{T^* - \varepsilon \, T^*}$$

where

$$\lim_{\varepsilon \to 0+} |S_{T^*}(u^2)|_{T^* - \varepsilon \, T^*} = 0.$$

This completes the proof.

As a consequence we obtain

Lemma 1.5.9: In the case (1.5.23) and under the assumption of Lemma 1.5.8 the condition (1.5.16) implies $v_{T(M)} \geq M$, if $T(M) > T_0$.

This Lemma follows from Lemma 1.5.8 and Theorem 1.5.6.

We shall see in Section 1.5.4 that the case (1.5.23) can be assumed in the problem of null-controllability of vibrations.

1.5.4 Applications to One-Dimensional Vibrations.

1.5.4.1 Distributed Control.

Here we give a slightly different formulation of the problem of null-controllability being defined in Section 1.1.1 from the one that was adopted in Section 1.4.1. If we replace, for a given

$T > 0$, the time variable $t \in [0, T]$ by $T - t$ in (1.1.1), (1.1.3) and (1.1.4) and define $\tilde{y}(\cdot, t) = y(\cdot, T - t)$, $\tilde{u}(t) = u(T - t)$, then (1.1.1), (1.1.3) hold as well for $\tilde{y}$ and $\tilde{u}$ and (1.1.4) reads

$$\tilde{y}(\cdot, T) = y_0, \quad \tilde{y}_t(\cdot, T) = -y_1 \quad \text{on} \quad (0, 1). \tag{1.1.}$$

For (1.1.5) we obtain

$$\tilde{y}(\cdot, 0, \tilde{u}) = \tilde{y}_t(\cdot, 0, \tilde{u}) = 0 \quad \text{a.e. on} \quad (0, 1). \tag{1.1.}$$

The problem of null-controllability for the given $T > 0$ now turn out to be equivalent to finding some $\tilde{u} \in L^2(0, T)$ such that the co responding solution $\tilde{y} = \tilde{y}(x, t, \tilde{u})$, $x \in [0, 1]$, $t \in [0, T]$ of (1.1.1), (1.1.3) for $y = \tilde{y}$ and of (1.1.5) satisfies (1.1.4). If one de fines $\tilde{S}_T : L^2(0, T) \to E \times L^2(0, 1)$ by

$$\tilde{S}_T(u) = \left(\sum_{j=1}^{\infty} \frac{h_j}{\sqrt{\lambda_j}} \int_0^T \tilde{u}(t) \sin\sqrt{\lambda_j}\, t \, dt \; e_j, \; \sum_{j=1}^{\infty} h_j \int_0^T \tilde{u}(t) \cos\sqrt{\lambda_j}\, t \, dt \; e_j \right)$$

$$\text{for } \tilde{u} \in L^2(0, T), \tag{1.4}$$

then (1.1.4) becomes equivalent with $\tilde{S}_T(\tilde{u}) = (y_0, -y_1)$ and the problem of null-controllability turns out to be equivalent wit the solvability of this equation for $y_0 \in E$ and $y_1 \in L^2(0, 1)$ being given.

For the beginning let $\hat{T} > 0$ be chosen arbitrarily and let $X = L^2(0, \hat{T})$ be equipped with the L^2-norm. For every $T \in [0, \hat{T}]$ the operator $\tilde{S}_T$ being defined by (1.4.1) defines a continuous line mapping from X into $Y = E \times L^2(0, 1)$ equipped with the norm

$$\|(y_1, y_2)\|_Y = \left(\|y_1\|_E^2 + \|y_2\|_{L^2(0,1)}^2 \right)^{1/2},$$

$y_1 \in E$, $y_2 \in L^2(0, 1)$ with $\|\cdot\|_E$ defined by (1.1.10). The function $y = y(T)$, $T \in [0, \hat{T}]$, in (1.5.2) is given by $y(T) = y = \begin{pmatrix} y_0 \\ -y_1 \end{pmatrix}$ fo all $T \in [0, \hat{T}]$. We assume $y \neq \Theta_Y$.

At first we will check the assumptions 1) - 3) of Theorem 1.5 Assumption 4) is trivially satisfied by the assumption $y \neq \Theta_Y$ By the definition (1.4.1) of $\tilde{S}_T$ the assumption 1) clearly hol In order to verify 2) we take $0 \leq T^* < T \leq \hat{T}$. Then, for every $u \in$

As a consequence of Theorems 1.5.4 and 1.5.6 we obtain

Theorem 1.5.7: Let the assumptions 1) - 4) of Theorem 1.5.1
hold together with the above assumptions 5), 6) and (1.5.16).
If $T(M) > T_0$, then

a) $\quad V_{T(M)} = M.$ $\hfill (1.5.20)$

b) Each control $u_M \epsilon U_M$ with (1.5.4) satisfies

$$S^*_{T(M)} (y^*)(u_M) = |S^*_{T(M)} (y^*)|.M \hfill (1.5.21)$$

for some $y^* \epsilon W^*$ with $S^*_{T(M)} (y^*) \neq \Theta_{X^*}$ which is independent of u_M
where W^* is the dual space of $W = S_{T(M)} (X)$ being equipped with
the norm $|y|_W = |y|_{\wedge}$, $y \epsilon W$, and $S^*_{T(M)}$ is the adjoint operator of
$S_{T(M)}$ considered as a continuous linear mapping from X onto W
with respect to the norm $|.|_W$.

Remark: The statement (1.5.21) is equivalent with the
maximum-principle

$$S^*_{T(M)} (y^*)(u_M) = \sup\{S^*_{T(M)} (y^*)(u) \mid u \epsilon U_M\}. \hfill (1.5.22)$$

For the remainder of this section we assume the function
$y : [0,\hat{T}] \to Y$ to be constant, in particular,

$$y(T) = \hat{y} \epsilon W, \; \hat{y} \neq \Theta_W, \quad \text{for all } T \epsilon [0,\hat{T}]. \hfill (1.5.23)$$

Then the first part of the above assumption 6) is trivially
satisfied. The second part is ensured by

Lemma 1.5.8: For each pair T_1, T_2 with $0 < T_1 \le T_2 \le \hat{T}$ and each $u \epsilon X$
we assume the existence of two elements $u^1_{T_1}$, $u^2_{T_1} \epsilon X$ such that

$$S_{T_2} (u) = S_{T_2} (u^1_{T_1}) + S_{T_2} (u^2_{T_1}), \; |u^1_{T_1}|_X \le |u|_X,$$

and $\hfill (1.5.24)$

$$\lim_{T_1 \to T_2 - 0} |S_{T_2} (u^2_{T_1})|_{T_2} = 0.$$

Then (1.5.19) is satisfied for every $T^* \epsilon (T_0, \hat{T}]$.

Proof: Let $\varepsilon_0 > 0$ be given such that $T_0 < T^* - \varepsilon_0$. Then, for every $\varepsilon \in (0, \varepsilon_0]$ and every $y \in W$, we have

$$|y|_{T^*} \leq |y|_{T^* - \varepsilon} \leq |y|_{T^* - \varepsilon_0} \leq \lambda(T^* - \varepsilon_0, T^*)|y|_{T^*} \quad \text{for some}$$

$\lambda(T^* - \varepsilon_0, T^*) > 0$ which is independent of y. Let us choose $u \in X$ with $S_{T^*}(u) = y$ and $|u|_X = |y|_{T^*}$. Then, for $T_2 = T^*$ and $T_1 = T^* - \varepsilon$, it follows from (1.5.24) that

$$|y|_{T^* - \varepsilon} \leq |S_{T^*}(u^1)|_{T^* - \varepsilon} \, T^* - \varepsilon + |S_{T^*}(u^2)|_{T^* - \varepsilon} \, T^* - \varepsilon$$

$$\leq |u^1|_{T^* - \varepsilon} \, X + \lambda(T^* - \varepsilon_0, T^*)|S_{T^*}(u^2)|_{T^* - \varepsilon} \, T^*$$

$$\leq |u|_X + \lambda(T^* - \varepsilon_0, T^*)|S_{T^*}(u^2)|_{T^* - \varepsilon} \, T^*$$

$$= |y|_{T^*} + \lambda(T^* - \varepsilon_0, T^*)|S_{T^*}(u^2)|_{T^* - \varepsilon} \, T^*$$

Therefore

$$0 \leq |y|_{T^* - \varepsilon} - |y|_{T^*} \leq \lambda(T^* - \varepsilon_0 T^*)|S_{T^*}(u^2)|_{T^* - \varepsilon} \, T^*$$

where

$$\lim_{\varepsilon \to 0+} |S_{T^*}(u^2)|_{T^* - \varepsilon} \, T^* = 0.$$

This completes the proof.

As a consequence we obtain

Lemma 1.5.9: In the case (1.5.23) and under the assumption of Lemma 1.5.8 the condition (1.5.16) implies $v_{T(M)} \geq M$, if $T(M) > T_0$.

This Lemma follows from Lemma 1.5.8 and Theorem 1.5.6.

We shall see in Section 1.5.4 that the case (1.5.23) can be assumed in the problem of null-controllability of vibrations.

1.5.4 Applications to One-Dimensional Vibrations.

1.5.4.1 Distributed Control.

Here we give a slightly different formulation of the problem of null-controllability being defined in Section 1.1.1 from the one that was adopted in Section 1.4.1. If we replace, for a given

$T > 0$, the time variable $t \in [0,T]$ by $T - t$ in (1.1.1), (1.1.3) and (1.1.4) and define $\tilde{y}(\cdot,t) = y(\cdot,T-t)$, $\tilde{u}(t) = u(T-t)$, then (1.1.1), (1.1.3) hold as well for $\tilde{y}$ and $\tilde{u}$ and (1.1.4) reads

$$\tilde{y}(\cdot,T) = y_0, \quad \tilde{y}_t(\cdot,T) = -y_1 \quad \text{on} \quad (0,1). \tag{1.1.4}$$

For (1.1.5) we obtain

$$\tilde{y}(\cdot,0,\tilde{u}) = \tilde{y}_t(\cdot,0,\tilde{u}) = 0 \quad \text{a.e. on} \quad (0,1). \tag{1.1.5}$$

The problem of null-controllability for the given $T > 0$ now turns out to be equivalent to finding some $\tilde{u} \in L^2(0,T)$ such that the corresponding solution $\tilde{y} = \tilde{y}(x,t,\tilde{u})$, $x \in [0,1]$, $t \in [0,T]$ of (1.1.1), (1.1.3) for $y = \tilde{y}$ and of (1.1.5) satisfies (1.1.4). If one defines $\tilde{S}_T : L^2(0,T) \to E \times L^2(0,1)$ by

$$\tilde{S}_T(u) = (\sum_{j=1}^{\infty} \frac{h_j}{\sqrt{\lambda_j}} \int_0^T \tilde{u}(t)\sin\sqrt{\lambda_j}t \; dt \; e_j, \; \sum_{j=1}^{\infty} h_j \int_0^T \tilde{u}(t)\cos\sqrt{\lambda_j}t \; dt \; e_j)$$

$$\text{for } \tilde{u} \in L^2(0,T), \tag{1.4.1}$$

then (1.1.4) becomes equivalent with $\tilde{S}_T(\tilde{u}) = (y_0,-y_1)$ and the problem of null-controllability turns out to be equivalent with the solvability of this equation for $y_0 \in E$ and $y_1 \in L^2(0,1)$ being given.

For the beginning let $\hat{T} > 0$ be chosen arbitrarily and let $X = L^2(0,\hat{T})$ be equipped with the L^2-norm. For every $T \in [0,\hat{T}]$ the operator $\tilde{S}_T$ being defined by (1.4.1) defines a continuous linear mapping from X into $Y = E \times L^2(0,1)$ equipped with the norm

$$\|(y_1,y_2)\|_Y = (\|y_1\|_E^2 + \|y_2\|_{L^2(0,1)}^2)^{1/2},$$

$y_1 \in E$, $y_2 \in L^2(0,1)$ with $\|\cdot\|_E$ defined by (1.1.10). The function $y = y(T)$, $T \in [0,\hat{T}]$, in (1.5.2) is given by $y(T) = y = \begin{pmatrix} y_0 \\ -y_1 \end{pmatrix}$ for all $T \in [0,\hat{T}]$. We assume $y \neq \Theta_Y$.

At first we will check the assumptions 1) - 3) of Theorem 1.5.1. Assumption 4) is trivially satisfied by the assumption $y \neq \Theta_Y$. By the definition (1.4.1) of $\tilde{S}_T$ the assumption 1) clearly holds. In order to verify 2) we take $0 \leq T^* < T \leq \hat{T}$. Then, for every $u \in X$, we obtain

$$| \tilde{S}_T(u) - \tilde{S}_{T^*}(u) |_Y^2 \leq (\sum_{j=1}^{\infty} h_j^2) (T-T^*) |u|_X^2$$

which implies

$$| \tilde{S}_T - \tilde{S}_{T^*} | \leq (\sum_{j=1}^{\infty} h_j^2)^{1/2} \sqrt{T-T^*}$$

and thus guarantees 2).

As to assumption 3) we first observe that X is a separable Hilbert space and thereby its own dual space. Weak[*] convergence in X is equivalent to weak convergence. In order to verify 3) it suffices to show that S_T (1.4.1) is compact for every $T \in [0, \hat{T}]$. For this purpose we define, for every $N \in \mathbb{N}$,

$$\tilde{S}_T^N(u) = \begin{pmatrix} \sum_{j=1}^{N} \frac{h_j}{\sqrt{\lambda_j}} \int_0^T u(t) \sin \sqrt{\lambda_j}\, t \, dt \, e_j \\[2mm] \sum_{j=1}^{N} h_j \int_0^T u(t) \cos \sqrt{\lambda_j}\, t \, dt \, e_j \end{pmatrix} \qquad (1.4.1)_N$$

$u \in X$. Since $\tilde{S}_T^N$ has a finite-dimensional range in Y, it is compact for every N. Now, for every $u \in X$, we have

$$| \tilde{S}_T(u) - \tilde{S}_T^N(u) |_Y^2 \leq (\sum_{j=N+1}^{\infty} h_j^2) T |u|_X^2$$

which implies

$$| \tilde{S}_T - \tilde{S}_T^N | \leq (\sum_{j=N+1}^{\infty} h_j^2)^{1/2} T^{1/2} \to 0 \quad \text{as } N \to \infty.$$

This together with the compactness of every $\tilde{S}_T^N$ ensures the compactness of $\tilde{S}_T$ and thus the assumption 3).

Before we can apply Theorem 1.5.1 we have to show that the minimum times T(M) defined by (1.1.7) and (1.5.3) coincide. In view of the above considerations it suffices to prove that the following two sets A and B defined by

$A = \{ T \in [0, \hat{T}] \mid \tilde{S}_T(u) = y \text{ for some } u \in U_M \ (1.5.1) \}$

and

$B = \{ T \in [0, \hat{T}] \mid \tilde{S}_T(u) = y \text{ for some } u \in L^2(0,T)$

$$\text{with } |u|_{L^2(0,T)} \leq M \},$$

respectively, are equal which can be easily seen. From
Theorem 1.5.1 we then infer the existence of a control
$u_M \in U_M$ with $\tilde{S}_{T(M)}(u_M) = y$ and $T(M) = \inf\{T \mid T \in A\} = \inf\{T \mid T \in B\}$.

For the following we make the assumption 5) at the beginning
of Section 1.5.3.

By the results of Section 1.4.1 it is, for instance, satisfied,
if (1.4.15) holds, for some $T \in (0, \hat{T}]$ in which case T_0 can be
chosen as

$$T_0 = \sup\left\{\frac{2\pi}{\sqrt{\bar{\lambda}_j} - \sqrt{\bar{\lambda}_{j-1}}} \mid j \in \mathbb{N} \cup \{0\}\right\}, \tag{1.5.25}$$

and if W is chosen as $E_r \times F_r$ with E_r and F_r being defined by
(1.4.4) and (1.4.5), respectively. The assumption (1.5.16) is
an immediate consequence of

$$\{u \in L^2(0,\hat{T}) \mid S_{T_1}(u) = y\} \subseteq \{u \in L^2(0,\hat{T}) \mid S_{T_2}(u) = y\},$$

if $0 < T_1 \leq T_2 \leq \hat{T}$. From the results of Section 1.5.3 (see in particular
Lemma 1.5.9) we then conclude $v_{T(M)} = M$, if $T(M) > T_0$ and if we
can verify the assumption of Lemma 1.5.8. For this purpose let
$T_0 < T_1 \leq T_2 \leq T$ and some $u \in X$ be given. The we define $u^1_{T_1}$, $u^2_{T_1} \in X$ by

$$u^1_{T_1} = \begin{cases} u & \text{a.e. on } [0, T_1] \\ 0 & \text{a.e. on } (T_1, \hat{T}], \end{cases} \quad , \quad u^2_{T_1} = \begin{cases} 0 & \text{a.e. on } [0, T_1], \\ u & \text{a.e. on } (T_1, \hat{T}]. \end{cases}$$

Then

$$S_{T_2}(u) = S_{T_2}(u^1_{T_1} + u^2_{T_1}) = S_{T_2}(u^1_{T_1}) + S_{T_2}(u^2_{T_1}) \text{ and}$$

$|u^1_{T_1}|_X \leq |u|_X$. If we put $y_{T_1} = S_{T_2}(u^2_{T_1})$, then

$$|S_{T_2}(u^2_{T_1})|_{T_2} = v_{T_2}(y_{T_1}) \leq |u^2_{T_1}|_{L^2(0,T_2)} \to 0 \text{ as } T_1 \to T_2 - 0.$$

Thus (1.5.24) holds and implies $v_{T(M)} = M$ as mentioned above.
Furthermore, we have

$$|u_M|_X = v_{T(M)} = M \tag{1.5.26}$$

for every time-minimal control u_M which is only possible, if $u_M = 0$ a.e. on $(T(M), \hat{T}]$. Since X is a Hilbert space, statement b) of Theorem 1.5.7 is essentially equivalent with (1.5.26) whose intuitive meaning is that all resources have to be exploited, if the state of rest of the vibrating system is to be achieved in minimum time.

1.5.4.2. Boundary Control.

In this case we can also proceed as in Section 1.5.4.1 and by reversing the time direction reduce the problem of null-controllability for some time $T \epsilon (0, \hat{T}]$, $\hat{T} > 0$ given arbitrarily, to the solvability of an operator equation of the form $S_T(v) = y$ for some fixed $y \epsilon E \times L^2(0,1)$ and S_T being defined by $S_T(v) = S(v)$ (1.4.16), $v \epsilon H_0^2(0,T)$. We could then obtain a family of continuous linear mappings from $X = H_0^2(0,\hat{T})$ into $Y = E \times L^2(0,1)$ with the properties 1) - 3) which are sufficient to prove Theorem 1.5.1 which means to guarantee the existence of time-minimal controls when restricted null-controllability for $T = \hat{T}$ is possible.

Within this model it is, however, not possible to verify the assumption of Lemma 1.5.8 which was used in Section 1.5.4.1 for the proof of (1.5.26), if $T(M) > T_0$ (see condition 5) at the beginning of Section 1.5.3). We therefore replace, as in Section 1.1.3, the problem of null-controllability by the solvability of the moment problem (1.1.23), (1.1.44), (1.1.45). So again we choose $\hat{T} > 0$ arbitrarily and define $X = L^2(0,\hat{T})$ equipped with the L^2-norm. We assume that there exists some $T_0 \epsilon [0,\hat{T})$ such that, for every $T \epsilon (T_0, \hat{T}]$, by

$$S_T(u) = \begin{pmatrix} \int_0^T tu(t) \, dt \\ \int_0^T u(t) \, dt \\ \int_0^T u(t) \cos \sqrt{\bar{\lambda}_j} t \, dt \\ \int_0^T u(t) \sin \sqrt{\bar{\lambda}_j} t \, dt \end{pmatrix}_{j \in \mathbb{N}} , \quad u \epsilon X ,$$

a linear mapping from X onto l^2 (over $\mathbb{R}$) is defined.

This is the case, if for some $T \in (0,\hat{T}]$ the condition (1.2.53) is satisfied for some $\delta > 0$ (or (1.4.15) for some $\varepsilon > 0$) and if T_0 is defined by (1.5.25). Because then, by Theorem 1.2.21, it follows that $1^2 \subseteq S_T(X)$. Conversely, if $c = S_T(u)$ for some $u \in X$, then for $z_0(t) = t$, $z_1(t) = 1$, $z_{2j}(t) = \cos\sqrt{\lambda_j}\,t$, $z_{2j+1}(t) = \sin\sqrt{\lambda_j}\,t$, $j \in \mathbb{N}$, $t \in [0,\hat{T}]$, it follows from Lemma 1.4.4 that

$$\sum_{j,k=2}^{N} <z_j,z_k>_{L^2(0,T)}\, a_j a_k \le \frac{3\pi\sqrt{2}}{T} \sum_{j=1}^{M} |a_j|^2$$

for all $N \ge 2$ and all $(a_2,\ldots,a_N) \in \mathbb{R}^{N-1}$ and in turn (see Theorem 1.2.1 and the following remark)

$$\|u\|_{L^2(0,T)} \ge \frac{T}{3\pi\sqrt{2}} \sum_{j=2}^{N} |c_j|^2$$

which implies $c \in 1^2$, hence $S_T(X) \subseteq 1^2$.

Now it can be seen as in Section 1.5.4.1 that the assumption of Lemma 1.5.8 is satisfied and, consequently, (1.5.26) holds, if $T(M) > T_0$.

1.6. Bibliographical Remarks and References.

The theory of exact solvability of linear operator equations which was developed in Section 1.3.1 has been adopted from the book [1] by S. Goldberg. The main result is Theorem 1.3.4 which plays a key role in the proof of null-controllability via operator equations.

In the case of distributed control the linear operator under consideration is given by (1.4.1) and is continuous. Therefore Theorem 1.3.8 can be applied in order to prove null-controllability. Theorem 1.3.8 is a consequence of Theorem 1.3.5 which is a special case of Theorems 1.3.3 and 1.3.4.

In the case of boundary control the linear operator S under consideration is defined by (1.4.16) and is not continuous. With the aid of Lemma 1.3.1, however, it can be shown that S is closed and thus Theorem 1.3.4 is applicable. The results on boundary null-controllability in Section 1.4.2 have also been derived in [2].

The general maximum-principle for minimum norm controls and the reduction of time-minimal controllability to norm-minimal controllability proved in Section 1.5.2 and 1.5.3, respectively, have been taken from [3].

References

[1] Goldberg, S.: Unbounded Linear Operators. New York - St. Louis -
     San Francisco - Toronto - London - Sydney: McGraw-Hill Book
     Company 1966.

[2] Krabs, W.: Remarks on Null-Controllability of One-Dimensional
     Vibrating Systems. In: Hoffmann, K.-H., and Krabs, W.: Optimal
     Control of Partial Differential Equations. Basel - Boston -
     Stuttgart: Birkhäuser Verlag 1984.

[3] Krabs, W., and Schmidt, E.J.P.G.: Time Minimal Controllability
     of Linear Systems. In: Mathematical Methods in Operations Re-
     search. Sofia 1981.

[4] Rolewicz, S.: Funktionalanalysis und Steuerungstheorie. Berlin -
     Heidelberg - New York: Springer-Verlag 1976.

2. Optimal Control of Heating Processes.
2.1. Problems in One Space Dimension.
2.1.1. Distributed Control.

The subject of the following investigations will be a heating
process in a one-dimensional medium which is given by the
interval [0,1] and whose temperature $y = y(x,t)$ as a function
of the space variable $x \in [0,1]$ and the time $t \in [0,\infty)$ develops
according to a linear parabolic partial differential equation
of the form

$$y_t(x,t) = \frac{\partial}{\partial x} (p(x)y_x(x,t)) + q(x) y(x,t) + r(x) u(t),$$

$$x \in (0,1), \quad t \in (0,\infty) \tag{2.1.1}$$

and boundary conditions of the form

$$a_0 y(0,t) + b_0 y_x(0,t) = 0,$$

$$a_1 y(1,t) + b_1 y_x(1,t) = 0, \quad t \in [0,\infty) \tag{2.1.2}$$

with $a_0^2 + b_0^2 > 0$ and $a_1^2 + b_1^2 > 0$. Here p and q are C^∞-functions
on [0,1] with p being positive and r belongs to the Hilbert space
$L^2(0,1)$. All three functions are chosen to be fixed during the
process whereas u is a control function which is variably chosen
in $L^\infty(0,\infty)$ and has to influence the heating process which is
assumed to start with an initial state at $t = 0$ given by

$$y(\cdot,0) = y_0 \quad \text{a.e. in } (0,1) \tag{2.1.3}$$

where $y_0 \in L^2(0,1)$ is also chosen to be fixed.

Every control function $u \in L^\infty(0,\infty)$ is considered as an instrument
to change the temperature of the process as to achieve certain
goals. We will consider some of these. The first can be expressed as

Problem of Controllability: Let some time T>0 and some final state
$y_T \in L^2(0,1)$ be given. Does there exist a control function $u \in L^\infty(0,\infty)$
which transfers the initial state $y_0 \in L^2(0,1)$ at $t = 0$ to the
target state $y_T \in L^2(0,1)$ at $t = T$?

This means that there exists a solution $y = y(x,t)$, $x \in [0,1]$,
$t \in [0,T]$ of the initial-boundary-value problem (2.1.1), (2.1.2),
(2.1.3) with

$$y(\cdot,T) = y_T \quad \text{a.e. in } (0,1). \tag{2.1.4}$$

If $y_0 = 0$, then this problem is also termed as <u>problem of null-reachability</u>, and if $y_T = 0$, it is called the <u>problem of null-controllability</u>. If in addition $u \in L^\infty(0,\infty)$ is required to satisfy

$$|u| \leq M \quad \text{a.e. on } (0,\infty) \tag{2.1.5}$$

for some constant $M>0$, then one speaks of the <u>problem of restricted controllability</u>. We shall see later that, in general, controllability holds for every $T>0$, if it holds for some $T>0$. So in this case the question arises whether restricted controllability is possible for some $T>0$. This question leads further to the

Problem of Time-Minimal Controllability:

Assume that restricted controllability for some $T>0$ and $y_T \in L^2(0,1)$ is possible. Then the least time $T(M)$ being the infimum of all times $T>0$ for which restricted controllability is possible is well defined and the question arises whether restricted controllability is possible for $T = T(M)$.

This question can be answered affirmatively by routine compactness arguments as we shall see later.

Far more difficult to answer is the question under which conditions time-minimal control functions $u \in L^\infty(0,\infty)$ have the so called "<u>bang-bang-property</u>" on $[0,T(M)]$, i.e.

$$|u| = M \quad \text{a.e. on } [0,T(M)]. \tag{2.1.6}$$

This property expresses the intuitive idea that in order to reach y_T in the least possible time $T(M)$ all the controlling resources have to be exploited. In addition it leads to the statement that the restriction of time-minimal control functions to the minimum time interval $[0,T(M)]$ is unique which is a simple consequence of the fact that the set of time-minimal control functions is convex.

Before dealing with the problems of controllability and time-minimal controllability we have first to clarify in what sense the initial-boundary-value problem (2.1.1), (2.1.2), (2.1.3) has a solution, if $u \in L^\infty(0,\infty)$ is chosen.

We begin with the following

Definition: A function $y : [0,1] \times [0,\infty) \rightarrow \mathbb{R}$ is a generalized solution of (2.1.1), (2.1.2), (2.1.3) for some given $u \in L^\infty(0,\infty)$, if it has the following properties:

(i) For each $t \in [0,\infty)$ the function $y(\cdot,t)$ is in $L^2(0,1)$, for each $t \in (0,\infty)$ the derivatives $y_x(\cdot,t)$ and $y_{xx}(\cdot,t)$ in the sense of distributions belong to $L^2(o,1)$ (this implies $y(\cdot,t)$, $y_x(\cdot,t) \in C[0,1]$ for all $t \in (0,\infty)$).

(ii) The function $t \rightarrow y(\cdot,t)$ from $[0,\infty)$ into $L^2(0,1)$ is continuous and continuously differentiable with respect to the norm of $L^2(0,1)$ and (2.1.3) is satisfied in the sense that

$$\lim_{t \rightarrow 0+} \| y(\cdot,t) - y_0 \|_{L^2(0,1)} = 0. \tag{2.1.7}$$

(iii) For each $t \in (0,\infty)$ the equation (2.1.1) is satisfied for almost all $x \in (0,1)$ with y_x, y_{xx} in the sense of (i) and y_t in the sense of (ii).

(iv) The boundary conditions (2.1.2) are satisfied for all $t \in (0,\infty)$ (this makes sense because of $y(\cdot,t)$, $y_x(\cdot,t) \in C[0,1]$ for all $t \in (0,\infty)$ - see (i)).

From these properties an explicit representation of a generalized solution of (2.1.1), (2.1.2), (2.1.3) can be derived. For this purpose we consider first the linear differential operator L being defined by

$$(Lz)(x) = \frac{d}{dx}(p(x)z'(x)) + q(x)z(x), \quad x \in (0,1)$$

on

$$D_L = \{z \in H^2(0,1) \mid a_0 z(0) + b_0 z'(0) = a_1 z(1) + b_1 z'(1) = 0\}.$$

It is well-known (see, for instance, [1]) that L is self-adjoint and possesses a sequence $(-\lambda_k)_{k \in \mathbb{N}}$ of simple eigenvalues satisfying

$$0 \leq \lambda_1 < \lambda_2 < \lambda_3 < \ldots < \lambda_n < \ldots$$

with

$$\lim_{k \rightarrow \infty} \lambda_k = \infty.$$

In fact, it is known that

$$\lambda_k = \frac{\pi^2}{p^2}(k+a)^2 + O(k) \quad \text{as } k \rightarrow \infty$$

for some real constant α and

$$P = \int_0^1 p(x)^{-1} \, dx.$$

The normalized eigenfunctions φ_k corresponding to the eigenvalues $-\lambda_k$ form an orthogonal basis for $L^2(0,1)$, i.e., each $z \in L^2(0,1)$ has an expansion

$$z(x) = \sum_{k=1}^{\infty} a_k \varphi_k(x) \text{ with } a_k = \int_0^1 z(x) \varphi_k(x) \, dx, \quad k \in \mathbb{N},$$

convergent in $L^2(0,1)$.

Now let $y = y(x,t)$, $x \in [0,1]$, $t \in [0,\infty)$ be a generalized solution of (2.1.1), (2.1.2), (2.1.3) in the above sense. Then, by property (i), for every $t \in [0,\infty)$ the function $y(\cdot,t) \in L^2(0,1)$ has an expansion of the form

$$y(x,t) = \sum_{k=1}^{\infty} c_k(t) \varphi_k(x) \text{ with } c_k(t) = \int_0^1 y(x,t) \varphi_k(x) \, dx, \quad k \in \mathbb{N},$$

convergent in $L^2(0,1)$.

Since $y_0 \in L^2(0,1)$, there is an expansion of the form

$$y_0(x) = \sum_{k=1}^{\infty} c_k \varphi_k(x) \text{ with } c_k = \int_0^1 y_0(x) \varphi_k(x) \, dx, \quad k \in \mathbb{N},$$

and (2.1.7) implies that

$$c_k(0) = c_k \quad \text{for all } k \in \mathbb{N}. \tag{2.1.8}$$

By inserting y into (2.1.1) we obtain by virtue of property (iii)

$$\sum_{k=1}^{\infty} (\dot{c}_k(t) + \lambda_k c_k(t) + d_k u(t)) \varphi_k = 0 \text{ a.e. on } (0,1)$$

for all $t \in (0,\infty)$ where

$$d_k = \int_0^1 r(x) \varphi_k(x) \, dx, \quad k \in \mathbb{N}.$$

This implies

$$\dot{c}_k(t) + \lambda_k c_k(t) + d_k u(t) = 0, \quad k \in \mathbb{N},$$
$$\text{for almost all } t \in (0,\infty) \tag{2.1.9}$$

as a consequence of the completeness of $(\varphi_k)_{k \in \mathbb{N}}$. For each $k \in \mathbb{N}$ the unique absolutely continuous solution of (2.1.8), (2.1.9) is given by

$$c_k(t) = e^{-\lambda_k t}(c_k + d_k \int_0^t e^{\lambda_k s} u(s) \, ds), \quad t \in [0, \infty).$$

Hence a generalized solution $y = y(x,t)$ of (2.1.1), (2.1.2), (2.1.3) for some $u \in L^\infty(0,\infty)$ in the above sense is neccessarily of the form

$$y(x,t) = \sum_{k=1}^\infty e^{-\lambda_k t}(c_k + d_k \int_0^t e^{\lambda_k s} u(s) \, ds) \varphi_k(x) \qquad (2.1.10)$$

for $x \in [0,1]$, $t \in [0,\infty)$ where

$$c_k = \int_0^1 y_0(x) \varphi_k(x) \, dx,$$

$$d_k = \int_0^1 r(x) \varphi_k(x) \, dx, \quad k \in \mathbb{N}. \qquad (2.1.11)$$

Conversely, if for some $u \in L^2(0,\infty)$ a function $y : [0,1] \times [0,\infty) \to \mathbb{R}$ is defined by (2.1.10), then one can show that y is a generalized solution of (2.1.1), (2.1.2), (2.1.3) in the above sense and therefore the only one.

If $y_T \in L^2(0,1)$ has an expansion

$$y_T = \sum_{k=1}^\infty e_k \varphi_k, \quad e_k = \int_0^1 y_T(x) \varphi_k(x) \, dx, \quad k \in \mathbb{N}, \qquad (2.1.12)$$

then the controllability condition (2.1.4) can be expressed equivalently in the form

$$e^{-\lambda_k T} c_k + \int_0^T e^{-\lambda_k(T-s)} u(s) \, ds \, d_k = e_k$$

$$\qquad (2.1.13)$$

for all $k \in \mathbb{N}$

If $r \in L^2(0,1)$ is such that all $d_k \neq 0$, $k \in \mathbb{N}$, then (2.1.13) can be written in the form

$$\int_0^T e^{-\lambda_k s} v(s) \, ds = \frac{1}{d_k}(e_k - c_k e^{-\lambda_k T})$$

$$\qquad (2.1.14)$$

for all $k \in \mathbb{N}$

with $v(s) = u(T-s)$, $s \in [0,T]$. This is a typical system of exponential moment equations.

2.1.2. Boundary Control.

Instead of (2.1.1) we consider now the homogeneous differential equation

$$y_t(x,t) = \frac{\partial}{\partial x}(p(x)y_x(x,t)) + q(x)y(x,t),$$

$$x\in(0,1), \quad t\in(0,\infty), \tag{2.1.1'}$$

and instead of (2.1.2) we assume inhomogeneous boundary conditions of the form

$$a_0 y(0,t) + b_0 y_x(0,t) = (1-\sigma)u(t)$$

$$a_1 y(1,t) + b_1 y_x(1,t) = \sigma u(t), \quad t\in(0,\infty), \tag{2.1.2'}$$

for $\sigma = 0$ or $\sigma = 1$. The assumptions on p, q, u, a_0, b_0, a_1, and b_1 are the same as in Section 2.1.1. The control is now applied to one of the boundary conditions (2.1.2'). Again an initial condition (2.1.3) with $y_0 \in L^2(0,1)$ is assumed and the problems of controllability, restricted controllability, and time-minimal controllability are formulated as in Section 2.1.1.

In order to clarify in what sense the initial-boundary-value problem (2.1.1'), (2.1.2'), (2.1.3) has a solution we assume first that u be in $W^{1,\infty}(0,\infty)$, i.e., u has a representation of the form

$$u(t) = u_0 + \int_0^t v(s)\, ds, \quad t\in[0,\infty),$$

for some $v\in L^\infty(0,\infty)$ and $u_0\in\mathbb{R}$.

We further assume that the boundary-value problem

$$Az(x) = \frac{d}{dx}(p(x)z'(x)) + q(x)z(x) = 0$$

$$\text{for } x\in(0,1), \tag{2.1.15}$$

$$a_0 z(0) + b_0 z'(0) = 1 - \sigma,$$

$$a_1 z(1) + b_1 z'(1) = \sigma \tag{2.1.16}$$

has exactly one solution $z = z_\sigma \in H^2(0,1)$ (which is equivalent to saying that the operator A does not have zero as an eigenvalue).

Then the initial-boundary-value problem

$$\tilde{y}_t(x,t) = \frac{\partial}{\partial x}(p(x)\tilde{y}_x(x,t)) + q(x)\tilde{y}(x,t)$$

$$- z_\sigma(x)v(t), \quad x\in(0,1), \quad t\in(0,\infty), \tag{2.1.17}$$

$$a_0\tilde{y}(0,t) + b_0\tilde{y}_x(0,t) = 0,$$

$$a_1\tilde{y}(1,t) + b_1\tilde{y}_x(1,t) = 0, \quad t\in(0,\infty) \tag{2.1.18}$$

$$\tilde{y}(x,0) = y_0(x) - u_0z_\sigma(x)$$

for almost all $x\in(0,1)$ $\tag{2.1.19}$

has exactly one generalized solution in the sense of Section 2.1.1 which is given by (see (2.1.10))

$$\tilde{y}(x,t) = \sum_{k=1}^{\infty} e^{-\lambda_k t}(<y_0-u_0z_\sigma,\varphi_k>$$

$$- <z_\sigma,\varphi_k> \int_0^t e^{\lambda_k s} v(s)\ ds)\varphi_k(x), \tag{2.1.20}$$

$x\in[0,1]$, $t\in[0,\infty)$, where

$$<f,g> = \int_0^1 f(x)g(x)dx$$

for any pair f, $g\in L^2(0,1)$.

If we define

$$y(x,t) = \tilde{y}(x,t) + z_\sigma(x)u(t), \tag{2.1.21}$$

then y satisfies (2.1.1'), (2.1.2'), (2.1.3). Conversely, if this is the case, then $\tilde{y}$ defined by (2.1.21) is the unique generalized solution of (2.1.17), (2.1.18), (2.1.19). Therefore we take y being defined by (2.1.21) as the unique solution of (2.1.1'), (2.1.2'), (2.1.3), if u is chosen in $W^{1,\infty}(0,\infty)$. On using (2.1.20) we obtain from (2.1.21) the representation

$$y(x,t) = \sum_{k=1}^{\infty} e^{-\lambda_k t}(<y_0,\varphi_k>$$

$$+ \lambda_k<z_\sigma,\varphi_k> \int_0^t e^{\lambda_k s} u(s)\ ds)\varphi_k(x), \tag{2.1.22}$$

$x\in[0,1]$, $t\in[0,\infty)$. This representation will also be taken as the unique solution of (2.1.1'), (2.1.2'), (2.1.3), if $u\in L^\infty(0,\infty)$.

On using the expansion (2.1.12) for the target state $y_T\in L^2(0,1)$ the controllability condition (2.1.4) can now be expressed in the form

$$e^{-\lambda_k T} <y_0,\varphi_k> + \lambda_k <z_\sigma,\varphi_k> \int_0^T e^{-\lambda_k(T-s)} u(s)\ ds = e_k \qquad (2.1.23)$$

for all $k \in \mathbb{N}$.

The equations (2.1.23) can also be written down without the use of the solution $z_\sigma \in H^2(0,1)$ of (2.1.15), (2.1.16), since one calculates in the case $\sigma = 0$

$$\lambda_k <z_0,\varphi_k> = \begin{cases} \dfrac{p(0)}{a_0}\ \varphi_k'(0), & \text{if } b_0 = 0, \\[2ex] -\dfrac{p(0)}{b_0}\ \varphi_k(0), & \text{if } b_0 \ne 0, \end{cases} \qquad (2.1.24)_0$$

and in the case $\sigma = 1$

$$\lambda_k <z_1,\varphi_k> = \begin{cases} -\dfrac{p(1)}{a_1}\ \varphi_k'(1), & \text{if } b_1 = 0, \\[2ex] \dfrac{p(1)}{b_1}\ \lambda_k(1), & \text{if } b_1 \ne 0. \end{cases} \qquad (2.1.24)_1$$

If all $\lambda_k <z_\sigma,\varphi_k> \ne 0$, $k \in \mathbb{N}$, then (2.1.23) can be written in the form

$$\int_0^T e^{-\lambda_k s} v(s)\ ds = \frac{1}{\lambda_k <z_\sigma,\varphi_k>}\ (e_k - <y_0,\varphi_k> e^{-\lambda_k T})$$

for all $k \in \mathbb{N}$

with $v(s) = u(T-s)$, $s \in [0,T]$, which again is a system of exponential moment equations.

2.2. On Moment Problems in Banach Spaces.
2.2.1. Problems in General Banach Spaces.

Let Z be a Banach space and let $(z_j)_{j \in \mathbb{N}}$ be a linearly independent sequence in Z, i.e., every finite subsequence $(z_j)_{j=1,\ldots,n}$, $n \in \mathbb{N}$, is linearly independent. Further, let $(g_j)_{j \in \mathbb{N}}$ be an arbitrary sequence of real numbers.

Then the moment problem consists of finding some $x \in X = Z^* = $ dual space of Z such that

$$x(z_j) = g_j \quad \text{for all } j \in \mathbb{N}. \qquad (2.2.1)$$

On using Hahn-Banach's theorem one can prove the following
existence statement.

Theorem 2.2.1: Let J be a nonempty subset of $\mathbb{N}$. If the moment
problem

$$x(z_j) = g_j \quad \text{for all } j \in J \qquad\qquad (2.2.1)_J$$

has a solution $x \in X$, then for every $\gamma \geq |x|$ where

$$|x| = \sup\{x(z) \mid z \in Z, |z| \leq 1\}$$

it follows that

$$\sum_{j \in J_f} g_j y_j \leq \gamma \left| \sum_{j \in J_f} g_j z_j \right| \quad \text{for all } y_j \in \mathbb{R},$$

$$\qquad\qquad (2.2.2)$$

$j \in J_f$, and all finite subsets J_f of J.

Conversely, if (2.2.2) is true, then there is a solution $x \in X$ of
$(2.2.1)_J$ with $|x| \leq \gamma$.

Proof: 1) Let $x \in X$ be a solution of $(2.2.1)_J$. Then for every finite
subset J_f of J and reals g_j, $j \in J_f$, it follows that

$$\sum_{j \in J_f} g_j y_j = \sum_{j \in J_f} x(z_j) y_j = x\left(\sum_{j \in J_f} y_j z_j \right)$$

$$\leq |x| \left| \sum_{j \in J_f} y_j z_j \right|.$$

Hence (2.2.2) is true for every $\gamma \geq |x|$. Conversely, if (2.2.2)
is true for some $\gamma \geq 0$, then we define, for every finite subset
J_f of J and all $y_j \in \mathbb{R}$, $j \in J_f$,

$$x\left(\sum_{j \in J_f} y_j z_j \right) = \sum_{j \in J_f} g_j y_j. \qquad\qquad (2.2.3)$$

By virtue of the linear independence of the sequence $(z_j)_{j \in \mathbb{N}}$ we
obtain, by (2.2.3), a well-defined linear functional on the
subspace V of Z which is spanned by $\{z_j \mid j \in J\}$. From (2.2.2)
and (2.2.3) it follows that

$$x\left(\sum_{j \in J_f} y_j z_j \right) \leq \gamma \left| \sum_{j \in J_f} y_j z_j \right| \quad \text{for all } y_j \in \mathbb{R},$$

$j \in J_f$, and all finite subsets J_f of J.

Hence x is continuous on V with $|x| \leq \gamma$. By Hahn-Banach's theorem x can be extended to some $x \in X$ which has the same norm. The system $(2.2.1)_J$ is contained as a special case in the definition $(2.2.3)$ so that x defined by $(2.2.3)$ and extended to all of Z is a solution in X of $(2.2.1)_J$ with $|x| \leq \gamma$. This completes the proof.

Let in particular

$$J = J_N = \{1, \ldots, N\} \quad \text{for any } N \in \mathbb{N}.$$

Then we define

$$\gamma = \gamma_N = \sup\{ \sum_{j=1}^{N} g_j y_j \mid y_j \in \mathbb{R}, \; | \sum_{j=1}^{N} y_j z_j | \leq 1 \}. \qquad (2.2.4)$$

From this definition it follows that

$$\sum_{j=1}^{N} g_j y_j \leq \gamma \mid \sum_{j=1}^{N} y_j z_j | \quad \text{for all } (y_1, \ldots, y_N)^T \in \mathbb{R}^N$$

which is equivalent to $(2.2.2)$ for $J = J_N$.

This implies that, for every $N \in \mathbb{N}$, there is a solution $x = x_N \in X$ of

$$x(z_j) = g_j, \; j = 1, \ldots, N, \qquad (2.2.1)_N$$

with $|x_N| \leq \gamma_N$ and γ_N being defined by $(2.2.4)$. If $x \in X$ is any solution of $(2.2.1)_N$, then for every $(y_1, \ldots, y_N)^T \in \mathbb{R}^N$ with $| \sum_{j=1}^{N} y_j z_j | \leq 1$, we conclude that

$$\sum_{j=1}^{N} g_j y_j = \sum_{j=1}^{N} x(z_j) y_j = x(\sum_{j=1}^{N} y_j z_j) \leq |x|,$$

hence $\gamma_N \leq |x|$. As a result we therefore obtain

Theorem 2.2.2: For every $N \in \mathbb{N}$ there is a solution $x = x_N \in X$ of $(2.2.1)_N$ with least norm such that $|x_N| = \gamma_N$ with γ_N being defined by $(2.2.4)$.

Since the set

$$V_N = \{ (y_1, \ldots, y_N)^T \in \mathbb{R}^N \mid | \sum_{j=1}^{N} y_j z_j | \leq 1 \} \qquad (2.2.7)$$

is compact in $\mathbb{R}^N$ and the linear functional $(y_1,\ldots,y_N)^T \to$

$\sum\limits_{j=1}^N g_j y_j$ is continuous, there exists a vector $(y_1^N,\ldots,y_N^N)^T \in V_N$

such that

$$\sum_{j=1}^N y_j^N g_j = \gamma_N. \qquad (2.2.8)$$

Moreover, for every least norm solution $x = x_N \in X$ of $(2.2.1)_N$ and every $(y_1^N,\ldots,y_N^N)^T \in V_N$ with $(2.2.8)$ we obtain

$$x_N\left(\sum_{j=1}^N y_j^N z_j\right) = \sum_{j=1}^N y_j^N g_j = \gamma_N = |x_N|. \qquad (2.2.9)$$

Let us assume that

$$\gamma_\infty = \sup_{N \in \mathbb{N}} \gamma_N < \infty. \qquad (2.2.10)$$

Then $(2.2.2)$ is satisfied for $J = \mathbb{N}$ and $\gamma = \gamma_\infty$. Therefore Theorem 2.2.1 guarantees the existence of a solution $x = x_\infty \in X$ of $(2.2.1)$ with $|x_\infty| \leq \gamma_\infty$. If $x \in X$ is an arbitrary solution of $(2.2.1)$, then for every $N \in \mathbb{N}$ and every $(y_1,\ldots,y_N)^T \in V_N$ it follows that

$$\sum_{j=1}^N y_j g_j = \sum_{j=1}^N y_j x(z_j) = x\left(\sum_{j=1}^N y_j z_j\right) \leq |x|,$$

hence $\gamma_N \leq |x|$ which implies $\gamma_\infty \leq |x|$.

As a result we obtain

Theorem 2.2.3: Under the assumption $(2.2.10)$ (which is equivalent to $(2.2.2)$ for $J = \mathbb{N}$ and $\gamma = \gamma_\infty$) there is a least norm solution $x = x_\infty \in X$ of $(2.2.1)$ with $|x_\infty| = \gamma_\infty$.

Without further assumptions it is not possible to generalize $(2.2.9)$.

If Z is a separable Banach space one can proceed as follows in order to determine a least norm solution of $(2.2.1)$ in X, if $(2.2.10)$ is satisfied: For every $N \in \mathbb{N}$ one determines a least norm solution $x = x_N \in X$ of $(2.2.1)_N$. Since the sequence $(|x_N| = \gamma_N)_{N \in \mathbb{N}}$ is bounded by γ_∞, there is a subsequence $(x_{N_i})_{i \in \mathbb{N}}$ of $(x_N)_{N \in \mathbb{N}}$ which is weak* convergent to some $x_\infty \in X$ which solves $(2.2.1)$.

Since $x \to |x|$ is weak* lower semi-continuous on X it follows that $|x_\infty| \leq \lim\limits_{i \to \infty} \inf |x_{N_i}| \leq \gamma_\infty$ which implies that x_∞ is a least norm solution of (2.2.1) in X with $|x_\infty| = \gamma_\infty$ by the above arguments.

2.2.2. Connection with Equations for Nuclear Operators.

Let us assume that the sequence $(z_j)_{j \in \mathbb{N}}$ which appears in (2.2.1) satisfies the condition

$$\sum_{j=1}^{\infty} |z_j| < \infty. \tag{2.2.11}$$

If we then define

$$S(x) = \sum_{j=1}^{\infty} x(z_j) e_j, \quad x \in X, \tag{2.2.12}$$

with $e_j = (\delta_{ji})_{i \in \mathbb{N}}$, δ_{ji} = Kronecker's symbol, we obtain a continuous linear mapping S from X into ℓ^1 because of

$$|S(x)|_{\ell^1} = \sum_{j=1}^{\infty} |x(z_j)| \leq \sum_{j=1}^{\infty} |z_j| |x|.$$

This operator S is called a nuclear operator from X into ℓ^1 (see Section 1.3.3, in particular (1.3.13)). In order to determine the adjoint operator S^* from ℓ^∞ into X^* we take any sequence $y = (y_j)_{j \in \mathbb{N}} \in \ell^\infty$. Then, for every $x \in X$ it follows that

$$\sum_{j=1}^{\infty} y_j S(x)_j = \sum_{j=1}^{\infty} y_j x(z_j) = x(\sum_{j=1}^{\infty} y_j z_j)$$

because of

$$|\sum_{j=N}^{N+M} y_j z_j| \leq \sup_{j \in \mathbb{N}} |y_j| \sum_{j=N}^{N+M} |z_j|$$

for every pair N,M which shows that $(\sum_{j=1}^{N} y_j z_j)_{N \in \mathbb{N}}$ is a Cauchy sequence in Z converging to $\sum_{j=1}^{\infty} y_j z_j$. Therefore the adjoint operator S^* of S is given by

$$S^*(y) = \sum_{j=1}^{\infty} y_j z_j, \quad (y_j)_{j \in \mathbb{N}} \in \ell^\infty, \tag{2.2.13}$$

and maps ℓ^∞ continuously into $Z \subset X^*$. Let $g = (g_j)_{j \in \mathbb{N}} \in \ell^1$ be given. Then the moment equations (2.2.1) are equivalent with the operator equation

$$S(x) = g. \tag{2.2.14}$$

The quantity γ_∞ defined by (2.2.10) can also be expressed by

$$\gamma_\infty = \sup\{ \sum_{j=1}^\infty g_j y_j \mid (y_j)_{j \in \mathbb{N}} \in \ell^\infty, \; \mid \sum_{j=1}^\infty y_j z_j \mid \; \leq 1 \} \tag{2.2.15}$$

If $\gamma_\infty < \infty$, then, by Theorem 2.2.3, there exists a least norm solution $x_\infty \in X$ of (2.2.1) $\Longleftrightarrow$ (2.2.14) such that $\mid x_\infty \mid = \gamma_\infty$. In addition we have the

__Theorem 2.2.4:__ Let the mapping S (2.2.14) from X into ℓ^1 be surjective, i.e., let $S(X) = \ell^1$. Then, for every $g \in \ell^1$, there exists some $y^\infty \in \ell^\infty$ with $\mid \sum_{j=1}^\infty y_j^\infty z_j \mid \; \leq 1$ such that for each least norm solution $x^\infty \in X$ of (2.2.14) $\Longleftrightarrow$ (2.2.1) we have

$$x^\infty (\sum_{j=1}^\infty y_j^\infty z_j) = \mid x^\infty \mid \tag{2.2.16}$$

__Proof:__ By Theorem 1.3.5 there exists a constant $\lambda > 0$ such that

$$\mid y \mid_\infty = \sup_{j \in \mathbb{N}} \mid y_j \mid \; \leq \lambda \mid \sum_{j=1}^\infty y_j z_j \mid$$

for all $y = (y_j)_{j \in \mathbb{N}} \in \ell^\infty$. This implies that the set

$$C = \{ y = (y_j)_{j \in \mathbb{N}} \in \ell^\infty \mid \; \mid \sum_{j=1}^\infty y_j z_j \mid \; \leq 1 \}$$

is contained in the set

$$B_\lambda = \{ y = (y_j)_{j \in \mathbb{N}} \in \ell^\infty \mid \; \mid y \mid_\infty = \sup_{j \in \mathbb{N}} \mid y_j \mid \; \leq \lambda \}$$

which is weak* compact in ℓ^∞.

If we show that C is closed in ℓ^∞, then C is also weak* closed as it is convex. So let a sequence $(y^n)_{n \in \mathbb{N}}$ in C and some $y \in \ell^\infty$ be given with $\lim_{n \to \infty} \mid y^n - y \mid_\infty = 0$. Then

$$\left| \ \left| \sum_{j=1}^{\infty} y_j z_j \right| - \left| \sum_{j=1}^{\infty} y_j^n z_j \right| \ \right| \leq \left| \sum_{j=1}^{\infty} (y_j - y_j^n) z_j \right|$$

$$\leq \|y - y^n\|_{\infty} \sum_{j=1}^{\infty} |z_j| \to 0 \text{ as } n \to \infty,$$

which implies $\left| \sum_{j=1}^{\infty} y_j z_j \right| \leq 1$, i.e., $y \in C$.

From the weak* closedness of C and $C \subset B_\lambda$ we infer that C is weak* compact and, since $y \to \sum_{j=1}^{\infty} g_j y_j$ is a weak* continuous linear function from ℓ^∞ into $\mathbb{R}$, there exists some $y^\infty \in C$ with

$$\sum_{j=1}^{\infty} g_j y_j^\infty = \gamma_\infty .$$

Let $x^\infty \in X$ be any least norm solution of (2.2.14) $\iff$ (2.2.1), then

$$\|x^\infty\| = \gamma_\infty = \sum_{j=1}^{\infty} x^\infty (z_j) y_j^\infty = x^\infty \left(\sum_{j=1}^{\infty} y_j^\infty z_j \right)$$

which completes the proof.

Remark: If S is not surjective, but the image space $Y = S(X)$ is closed in ℓ^1 and hence also a Banach space with respect to the ℓ^1-norm, the assertion of Theorem 2.2.4 remains true for every $g \in Y$. In this case the dual space Y^* of Y can also be identified with ℓ^∞ and the adjoint operator S^* of S is also given by (2.2.13) so that the proof of Theorem 2.2.4 can be taken over without changes.

If $Y = S(X)$, however, is not closed, then Theorem 1.3.5 can no more be applied because it requires Y to be a Banach space. In this case we assume that Y can be equipped with a norm $\|\cdot\|_Y$ such that Y becomes a Banach space with respect to this norm and $S : X \to Y$ stays continuous. This is always possible (see Section 1.5.3) by defining

$$\|g\|_Y = \inf\{\|u\| \mid u \in X, S(u) = g\}, \ g \in Y.$$

The continuity of S then follows from

$$|S(u)|_Y = \inf\{|\tilde{u}| \mid \tilde{u} \in X, \ S(\tilde{u}) = S(u)\} \leq |u|.$$

From Theorem 1.5.4 (mutatis mutandis) we obtain

Theorem 2.2.5: Let $Y = S(X)$ be a Banach space with respect to some norm $|\cdot|_Y$ and let Y^* be the dual space of Y. Then, for every $g \in Y$ and every least norm solution $x = x^\infty$ of $(2.2.14)$ $\iff$ $(2.2.1)$ there exists some $y^* \in Y^*$ which is independent of x^∞ and satisfies

$$S^*(y^*)(x^\infty) = |x^\infty| \text{ and } |S^*(y^*)| = 1 \qquad (2.2.17)$$

where S^* denotes the adjoint operator of S and maps Y^* into X^*.

In general, it is difficult to determine X^*, Y^* and S^* so that the application of $(2.2.17)$ to special cases is more difficult than the application of $(2.2.16)$.

2.2.3. On Solving Finite Moment Problems.

We have seen in Section 2.2.1 how the infinite moment problem $(2.2.1)$ can be reduced to a sequence of finite moment problems $(2.2.1)_N$ for $N \in \mathbb{N}$. The question now arises how $(2.2.1)_N$ can be solved for a given $N \in \mathbb{N}$. We assume that $z_1, \ldots, z_N \in Z$ are linearly independent. By Theorem 2.2.2 there exists a least norm solution $x = x_N \in X$ of $(2.2.1)_N$. In order to determine x_N we assume $(g_1, \ldots, g_N) \neq \theta_N$ and define

$$H = \{(y_1, \ldots, y_N) \in \mathbb{R}^N \mid \sum_{j=1}^{N} g_j y_j = 1\} \qquad (2.2.18)$$

and consider the problem of finding $(\hat{y}_1, \ldots, \hat{y}_N) \in H$ such that

$$|\sum_{j=1}^{N} \hat{y}_j z_j| \leq |\sum_{j=1}^{N} y_j z_j| \text{ for all } (y_1, \ldots, y_N) \in H. \qquad (2.2.19)$$

The existence of $(\hat{y}_1, \ldots, \hat{y}_N) \in H$ with $(2.2.19)$ is ensured by well known arguments in approximation theory. If we put

$$H^0 = \{(y_1, \ldots, y_N) \in \mathbb{R}^N \mid \sum_{j=1}^{N} g_j y_j = 0\},$$

then

$$
| \sum_{j=1}^{N} \hat{y}_j z_j - \sum_{j=1}^{N} y_j^* z_j | \le | \sum_{j=1}^{N} \hat{y}_j z_j - \sum_{j=1}^{N} y_j z_j |
$$

$$
\text{for all } (y_1, \ldots, y_N) \in H^O.
$$

By a well known duality theorem in approximation theory there exists some $\hat{x} \in X$ with $|\hat{x}| = 1$,

$$
\hat{x} (\sum_{j=1}^{N} y_j z_j) = 0 \text{ for all } (y_1, \ldots, y_N) \in H^O \tag{2.2.20}
$$

and

$$
\hat{x} (\sum_{j=1}^{N} \hat{y}_j z_j) = | \sum_{j=1}^{N} \hat{y}_j z_j |. \tag{2.2.21}
$$

The statement (2.2.20) is equivalent to the existence of some $\rho \in \mathbb{R}$ such that

$$
\hat{x} (z_j) = \rho g_j, \quad j = 1, \ldots, N,
$$

which implies

$$
\rho = | \sum_{j=1}^{N} \hat{y}_j z_j | > 0 \tag{2.2.22}
$$

by virtue of (2.2.21). If we define $x_N = \frac{1}{\rho} \hat{x}$ and $y_j^N = \frac{1}{\rho} \hat{y}_j$ for $j = 1, \ldots, N$, then

$$
x_N (z_j) = g_j \text{ for } j = 1, \ldots, N
$$

and

$$
x_N (\sum_{j=1}^{N} y_j^N z_j) = | \sum_{j=1}^{N} \hat{y}_j z_j | / \rho^2 = \frac{1}{\rho} = | x_N |.
$$

Let $x \in X$ be any solution of $(2.2.1)_N$. Then

$$
|x| - |x_N| \ge x (\sum_{j=1}^{N} y_j^N z_j) - x_N (\sum_{j=1}^{N} y_j^N z_j)
$$

$$
= \sum_{j=1}^{N} y_j^N \underbrace{(x(z_j) - x_N(z_j))}_{= 0} = 0.
$$

This shows that x_N is a least norm solution of $(2.2.1)_N$.

Thus, in order to determine such x_N one has first to determine $(\hat{y}_1,\ldots,\hat{y}_N) \in H$ (2.2.18) which satisfies (2.2.19) and then to find some $\hat{x} \in X$ with $|\hat{x}| = 1$ which solves (2.2.20) and (2.2.21). Finally, one has to define $x_N = \frac{1}{\rho} \hat{x}$ with ρ given by (2.2.22). There are cases where $\hat{x}$ is uniquely defined by $|\hat{x}| = 1$ and (2.2.21) so that, by the above mentioned duality theorem $\hat{x}$ automatically satisfies (2.2.20). In such a case we are led to

Theorem 2.2.6: We assume that for every vector $y = (y_1,\ldots,y_N) \in \mathbb{R}^N$ with $y \neq \Theta_N$ there is exactly one $x \in X$ with $|x| = 1$ and

$$x\left(\sum_{j=1}^{N} y_j z_j\right) = \left| \sum_{j=1}^{N} y_j z_j \right|. \tag{2.2.23}$$

Then a least norm solution $x = x_N \in X$ is obtained by determining $(\hat{y}_1,\ldots,\hat{y}_N) \in H$ (2.2.18) with (2.2.19), by choosing the unique $\hat{x} \in X$ with $|\hat{x}| = 1$, (2.2.1) and by defining $x_N = \frac{1}{\rho} \hat{x}$ with ρ given by (2.2.22).

2.3. Equations for Nuclear Operators in Banach Spaces.
2.3.1 Exact Solvability.

Let Y and Z be Banach spaces and let $(y_j)_{j \in \mathbb{N}}$ and $(z_j)_{j \in \mathbb{N}}$ be sequences in Y and Z, respectively, such that

$$\sum_{j=1}^{\infty} |y_j| \, |z_j| < \infty. \tag{2.3.1}$$

Then

$$S(x) = \sum_{j=1}^{\infty} x(z_j) y_j, \quad x \in X = Z^*, \tag{2.3.2}$$

defines a linear mapping from X into Y. In order to see that $S(x) \in Y$ for each $x \in X$ we define, for every $r \in \mathbb{N}$,

$$S_n(x) = \sum_{j=1}^{n} x(z_j) y_j, \quad x \in X. \tag{2.3.1}_n$$

For any choice of $1 \leq n < m$ we then obtain

$$|S_m(x) - S_n(x)| \leq \left(\sum_{j=n+1}^{m} |z_j| \, |y_j| \right) |x|$$

which implies that $(S_n(x))_{n \in \mathbb{N}}$ is a Cauchy sequence in Y for every $x \in X$, by virtue of (2.3.1). Hence the completeness of Y implies the existence of

$$S(x) = \lim_{n \to \infty} S_n(x) \in Y.$$

The linearity of S is a consequence of the linearity of every S_n. Furthermore, it follows that

$$|S(x)| \leq \left(\sum_{j=1}^{\infty} |z_j| \, |y_j| \right) |x|$$

for all $x \in X$. Therefore (2.3.2) defines a continuous linear mapping from X into Y which is a generalization of the mapping (1.3.13) to Banach spaces and is also called a <u>nuclear operator</u>. If Y is a Hilbert space and $(y_j)_{j \in \mathbb{N}}$ is an orthonormal sequence in Y, then (2.3.1) implies

$$\sum_{j=1}^{\infty} |z_j|^2 < \infty \tag{2.3.3}$$

because the sequence $(|z_j|)_{j \in \mathbb{N}}$ is in ℓ^1 and hence also in ℓ^2. From (2.3.3) it follows, for each $x \in X$, that

$$\sum_{j=1}^{\infty} x(z_j)^2 \leq \left(\sum_{j=1}^{\infty} |z_j|^2 \right) |x|^2.$$

Hence, $S(x)$ defined by (2.3.2) is in Y and from

$$|S(x)|^2 = \sum_{j=1}^{\infty} x(z_j)^2 \leq \left(\sum_{j=1}^{\infty} |z_j|^2 \right) |x|^2, \quad x \in X$$

it follows that S defines a continuous linear mapping from X into Y.

If, in addition, Z is a Hilbert space, i.e., $X = Z^* = Z$, then (2.3.3) implies (1.3.10) with $\mu = \sum_{j=1}^{\infty} |z_j|^2$ and the nuclear operator (2.3.2) takes the form (1.3.13).

In order to determine the adjoint operator $S^* : Y^* \to X^*$ we again assume that Y and Z are Banach spaces and that (2.3.1) is satisfied.

Then, for each $y^* \in Y^*$, we obtain

$$y^*(S(x)) = \sum_{j=1}^{\infty} x(z_j) y^*(y_j), \quad x \in X.$$

For each $n \in \mathbb{N}$ we define

$$z^n_{y^*} = \sum_{j=1}^{n} y^*(y_j) z_j \in Z.$$

Then, for any choice of $1 \leq n < m$, it follows that

$$|z^m_{y^*} - z^n_{y^*}| \leq |y^*| \left(\sum_{j=n+1}^{m} |y_j| \, |z_j| \right)$$

which, by virtue of (2.3.1), implies that $(z^n_{y^*})_{n \in \mathbb{N}}$ is a Cauchy sequence in Z. Hence the completeness of Z implies the existence of

$$z_{y^*} = \sum_{j=1}^{\infty} y^*(y_j) z_j = \lim_{n \to \infty} z^n_{y^*} \in Z \qquad (2.3.4)$$

and z_{y^*} is uniquely defined by y^*. By the definition of S^* we conclude that, for each $y^* \in Y^*$,

$$S^*(y^*)(x) = y^*(S(x)) = \lim_{n \to \infty} x(z^n_{y^*}) = x(z_{y^*}) \qquad (2.3.5)$$

for all $x \in X$. Furthermore it follows that

$$|S^*(y^*)| = |z_{y^*}|. \qquad (2.3.6)$$

Summarizing we obtain the

Lemma 2.3.1: If Y and Z are Banach spaces and $(y_j)_{j \in \mathbb{N}}$ and $(z_j)_{j \in \mathbb{N}}$ are sequences in Y and Z, respectively, which satisfy (2.3.1) then by (2.3.2) a continuous linear mapping S from $X = Z^*$ into Y is defined whose adjoint mapping $S^* : Y^* \to X^*$ satisfies (2.3.5), (2.3.6) with z_{y^*} being defined by (2.3.4).

Remark: The assertions of Lemma 2.3.1 also hold true, if Y is a Hilbert space, Z is a Banach space, $(y_j)_{j \in \mathbb{N}}$ is an orthonormal sequence in Y, and $(z_j)_{j \in \mathbb{N}}$ is a sequence in Z which satisfies (2.3.3).

Under the assumptions of Lemma 2.3.1 we can immediately apply
Theorem 1.3.5 and obtain

Theorem 2.3.2: Under the assumptions of Lemma 2.3.1 the nuclear
operator S defined by (2.3.2) maps $X = Z^*$ onto Y, if and only if
there exists some constant $\lambda > 0$ such that

$$|y^*| \leq \lambda |\sum_{j=1}^{\infty} y^*(y_j)z_j| \text{ for all } y^* \in Y^*. \tag{2.3.7}$$

2.3.2. Approximate Solvability.

Instead of exactly solving equations of the form

$$S(x) = \sum_{j=1}^{\infty} x(z_j)y_j = y \tag{2.3.8}$$

for given elements $y \in Y$ we are now interested in the question of
approximate solvability of (2.3.8) for any $y \in Y$, i.e., in the
existence of a sequence $(x^n)_{n \in \mathbb{N}}$ in X with

$$\lim_{n \to \infty} S(x^n) = y. \text{ (see (1.3.8))}.$$

Globally this can be expressed by $\overline{S(X)} = Y$ where $\overline{A}$ denotes the
closure of a subset A of a normed linear space.

Approximate solvability of (2.3.8) is governed by Theorem 1.3.6
for $D(S) = X$ which states that it is equivalent to the injectivity
of the adjoint operator S^* which is given by (2.3.4), (2.3.5).
This in turn is related to the minimality of the sequence $(z_j)_{j \in \mathbb{N}}$
(see Lemma 1.3.9) which, by definition, requires that, for each
$k \in \mathbb{N}$, the element z_k does not belong to the closure of the span
of $\{z_j | j \in \mathbb{N}, j \neq k\}$ (see Section 1.2.1). As a slight generalization
of Lemma 1.3.9 we can prove

Lemma 2.3.3: In addition to (2.3.1) let the sequence $(z_j)_{j \in \mathbb{N}}$ in Z
be minimal and let the sequence $(y_j)_{j \in \mathbb{N}}$ in Y be complete which
means that $y^*(y_j) = 0$ for all $j \in \mathbb{N}$ and some $y^* \in Y^*$ implies $y^* = \Theta_{Y^*}$.
Then the adjoint operator S^* of the nuclear operator S (2.3.2) is
injective.

<u>Proof:</u> Assume $S^*(y^*) = \Theta_{X^*}$ for some $y^* \in Y^*$. Then, by (2.3.4), (2.3.5), it follows that

$$\sum_{j=1}^{\infty} y^*(y_j) z_j = \Theta_Z$$

and in turn

$$y^*(y_j) = 0 \text{ for all } j \in \mathbb{N} \implies y^* = \Theta_{Y^*}.$$

Otherwise, if $y^*(y_k) \neq 0$ for some $k \in \mathbb{N}$, we would have that

$$z_k = \sum_{\substack{j=1 \\ j \neq k}}^{\infty} - \frac{y^*(y_j)}{y^*(y_k)} z_j \in \overline{\text{span of } \{z_j \mid j \in \mathbb{N}, \ j \neq k\}}$$

which contradicts the minimality of the sequence $(z_j)_{j \in \mathbb{N}}$.

Summarizing we obtain the

<u>Theorem 2.3.4:</u> In addition to the assumptions of Lemma 2.3.1 let the sequence $(z_j)_{j \in \mathbb{N}}$ be minimal and let the sequence $(y_j)_{j \in \mathbb{N}}$ be complete. Then approximate solvability of (2.3.8) holds for every $y \in Y$, i.e., $\overline{S(X)} = Y$.

If Y is a Hilbert space and $(y_j)_{j \in \mathbb{N}}$ a complete orthonormal sequence in Y, then a direct proof of Theorem 2.3.4 can be given as in Section 1.3.3 for Theorem 1.3.10. Instead of Theorem 1.2.5 we now need

<u>Theorem 2.3.5:</u> A sequence $(z_j)_{j \in \mathbb{N}}$ is minimal in Z, if and only if there exists a sequence $(x_k)_{k \in \mathbb{N}}$ in $X = Z^*$ with

$$x_k(z_j) = \delta_{kj} \quad \text{for all } j, \ k \in \mathbb{N} \tag{2.3.9}$$

where δ_{kj} denotes Kronecker's symbol.

The proof is the same as that of Theorem 1.2.5.

The direct proof of Theorem 2.3.4 proceeds as follows: We choose any sequence $(x_k)_{k \in \mathbb{N}}$ in $X = Z^*$ with (2.3.9) and put

$$x^n = \sum_{j=1}^{n} \langle y, y_j \rangle x_j \quad \text{for every } n \in \mathbb{N}$$

where $y \in Y$ is given. From (2.3.9) it follows that

$$S(x^n) = \sum_{j=1}^{n} \langle y, y_j \rangle y_j$$

and

$$\lim_{n \to \infty} S(x^n) = \sum_{j=1}^{\infty} \langle y, y_j \rangle y_j = y,$$

by virtue of the completeness of $(y_j)_{j \in \mathbb{N}}$ in Y.

2.4. On Exponential Moment Problems

2.4.1. Exponential Sums and Series.

In this Section we collect a few results on exponential sums and series which are taken from L. Schwartz [11]. Proofs are partially omitted.

Let $(\lambda_j)_{j \in \mathbb{N}}$ be a strictly increasing sequence of positive real numbers. We consider the corresponding sequence

$$z_j(t) = e^{-2\pi \lambda_j t}, \quad j \in \mathbb{N}, \ t \in [0, \infty), \tag{2.4.1}$$

in $L_p[0, \infty)$ for any $p \in [1, \infty)$. We denote by $A_p(\Lambda)$ the closure of the linear subspace which is generated by the sequence (2.4.1) in $L_p[0, \infty)$.

As part of the Theorem of Müntz we have the following

Theorem 2.4.1: If

$$A = \sum_{j \in \mathbb{N}} \frac{1}{\lambda_j} < \infty, \tag{2.4.2}$$

then the sequence (2.4.1) is minimal in $L_p[0, \infty)$, i.e. (by Theorem 2.3.5) there exists a sequence $(x_j)_{j \in \mathbb{N}}$ in $L_q[0, \infty)$, $\frac{1}{p} + \frac{1}{q} = 1$, such that

$$\int_0^\infty x_j(t) z_k(t) \, dt = \delta_{jk} = \begin{cases} 0, & \text{if } j \neq k, \\ 1, & \text{if } j = k, \end{cases}$$

for all $j, k \in \mathbb{N}$.

In addition it can be shown that there exist constants $c_1, c_2, c_2 > 0$ such that

$$\|x_j\|_{L_q[0,\infty)} \le \frac{C_1}{\lambda_1 |\varphi(\lambda_j)|}, \text{ if } q \ge 2 (\implies 1 \le p \le 2), \qquad (2.4.3a)$$

and

$$\|x_j\|_{L_q[0,\infty)} \le \frac{C_2 A + C_3}{\lambda_1 |\varphi(\lambda_j)|}, \text{ if } 1 \le q \le 2 (\implies p \ge 2), \qquad (2.4.3b)$$

$$\text{for all } j \in \mathbb{N}$$

where

$$\varphi(\lambda_j) = \frac{1}{2\lambda_j (1+\lambda_j)^2} \prod_{\substack{k=1 \\ k \ne j}}^{\infty} \frac{\lambda_k - \lambda_j}{\lambda_k + \lambda_j}. \qquad (2.4.4)$$

Let $F \in A_p(\wedge)$ be given arbitrarily. Then F is of the form

$$F = \lim_{n \to \infty} \sum_{j=1}^{n} a_j^n z_j$$

where the convergence is in the L_p-sense. For each $j \in \mathbb{N}$ it then follows that

$$\int_0^{\infty} x_j(t) F(t) \, dt = \lim_{n \to \infty} a_j^n = a_j \qquad (2.4.5)$$

and by Hölder's inequality and (2.4.3) we obtain

$$|a_j| \le \|x_j\|_{L_q[0,\infty)} \|F\|_{L_p[0,\infty)} \le \frac{C}{|\varphi(\lambda_j)|} \|F\|_{L_p[0,\infty)} \qquad (2.4.6)$$

$$\text{for all } j \in \mathbb{N} \text{ and a suitable constant } C > 0.$$

This implies for each $n \in \mathbb{N}$ that

$$\sum_{j=1}^{n} |a_j| e^{2\pi(\lambda_1 - \lambda_j)t} \le C\|F\|_{L_p[0,\infty)} \sum_{j=1}^{n} \frac{e^{2\pi(\lambda_1 - \lambda_j)t}}{|\varphi(\lambda_j)|} \qquad (2.4.7)$$

where

$$\frac{1}{|\varphi(\lambda_j)|} \le \lambda_j (1+\lambda_j)^2 \frac{\prod_{k=1}^{\infty} (1 + \frac{\lambda_j}{\lambda_k})}{\prod_{\substack{k=1 \\ k \ne j}}^{\infty} 1 - \frac{\lambda_j}{\lambda_k}}. \qquad (2.4.8)$$

For the following we assume that there is some $\delta > 0$ such that

$$\lambda_{j+1} - \lambda_j \ge \delta \text{ for all } j \in \mathbb{N}$$

and $\qquad\qquad\qquad\qquad\qquad\qquad\qquad\qquad\qquad\qquad (2.4.9)$

$$\lambda_1 \ge \delta.$$

Let $\alpha > 0$ be given arbitrarily. Then there is some $n(\alpha) \in \mathbb{N}$ such that

$$\sum_{k \geq n(\alpha)} \frac{1}{\lambda_k} \leq \frac{\alpha}{4}$$

and from $1 + \xi \leq e^{\xi}$ for all $\xi \geq 0$ it follows that

$$\prod_{k \geq n(\alpha)} (1 + \frac{\lambda_j}{\lambda_k}) \leq \prod_{k \geq n(\alpha)} \exp(\frac{\lambda_j}{\lambda_k}) = \exp(\lambda_j \sum_{k \geq n(\alpha)} \frac{1}{\lambda_k}) \leq e^{\frac{\alpha}{4}\lambda_j},$$

consequently

$$\prod_{k=1}^{\infty} (1 + \frac{\lambda_j}{\lambda_k}) \leq \prod_{k=1}^{n(\alpha)-1} (1 + \frac{\lambda_j}{\lambda_k}) e^{\frac{\alpha}{4}\lambda_j} \quad \text{for all } j \in \mathbb{N}.$$

Since

$$\prod_{k=1}^{n(\alpha)-1} (1 + \frac{\lambda_j}{\lambda_k}) \leq (1 + \frac{\lambda_j}{\delta})^{n(\alpha)-1} \leq e^{\frac{\alpha}{4}\lambda_j} \quad \text{and}$$

$$\lambda_j (1 + \lambda_j)^2 \leq e^{\frac{\alpha}{4}\lambda_j} \quad \text{for all sufficiently large } j \in \mathbb{N}$$

it follows that there exists a constant $K_1(\alpha) > 0$ such that

$$\lambda_j (1 + \lambda_j)^2 \prod_{k=1}^{\infty} (1 + \frac{\lambda_j}{\lambda_k}) \leq K_1(\alpha) e^{\frac{3}{4}\alpha\lambda_j} \quad \text{for all } j \in \mathbb{N}. \tag{2.4.10}$$

By a Theorem of Hadamard it further follows from (2.4.9) that (see L. Schwartz [11], p. 31)

$$\prod_{\substack{k=1 \\ k \neq j}}^{\infty} |1 - \frac{\lambda_j}{\lambda_k}| \geq e^{-\frac{\alpha}{4}\lambda_j} \quad \text{for all sufficiently large } j \in \mathbb{N}.$$

Therefore there exists a constant $K_2(\alpha) > 0$ such that

$$\prod_{\substack{k=1 \\ k \neq j}}^{\infty} |1 - \frac{\lambda_j}{\lambda_k}|^{-1} \leq K_2(\alpha) e^{\frac{\alpha}{4}\lambda_j} \quad \text{for all } j \in \mathbb{N}.$$

This, in connection with (2.4.8) and (2.4.10), implies the existence of a constant $K(\alpha) > 0$ such that

$$\frac{1}{|\varphi(\lambda_j)|} \leq K(\alpha) e^{\alpha\lambda_j} \quad \text{for all } j \in \mathbb{N}. \tag{2.4.11}$$

For a given $\varepsilon>0$ we put $\alpha = \alpha(\varepsilon) = \varepsilon\pi\frac{\lambda_2-\lambda_1}{\lambda_2}$ and choose
$t \geq \varepsilon = \frac{\alpha\lambda_2}{\pi(\lambda_2-\lambda_1)}$.

Because of $\frac{\lambda_2}{\lambda_2-\lambda_1} = \frac{1}{1-\frac{\lambda_1}{\lambda_2}} \geq \frac{1}{1-\frac{\lambda_1}{\lambda_j}} = \frac{\lambda_j}{\lambda_j-\lambda_1}$ for all $j\geq 2$

which implies $\alpha\lambda_j + 2\pi(\lambda_1-\lambda_j)t \leq -\alpha\lambda_j$ for all $j\geq 2$ we deduce
from (2.4.7) and (2.4.11) that

$$\sum_{j=1}^{n} |a_j|e^{2\pi(\lambda_1-\lambda_j)t} \leq C|F|_{L_p[0,\infty)}K(\alpha)(e^{\alpha\lambda_1} + \sum_{j=2}^{\infty} e^{-\alpha\lambda_j})$$

and, because of $e^{-\alpha\lambda_j} \leq \frac{1}{1+\alpha\lambda_j} \leq \frac{1}{\alpha\lambda_j}$, that

$$\sum_{j=1}^{n} |a_j|e^{-2\pi\lambda_j t} \leq C|F|_{L_p[0,\infty)}K(\alpha)e^{-\lambda_1\varepsilon}(e^{\alpha\lambda_1} + \frac{1}{\alpha}\sum_{j=2}^{\infty}\frac{1}{\lambda_j}).$$

$$(2.4.12)$$

If we define $u = t + is$, $i = \sqrt{-1}$, $t>0$, $s\in\mathbb{R}$, and put

$$G(u) = \sum_{j=1}^{\infty} a_j e^{-2\pi\lambda_j u}, \qquad\qquad (2.4.13)$$

then by (2.4.12) the series (2.4.13) converges absolutely and
uniformly for all $u = t + is$ with $t\geq\varepsilon$ for any $\varepsilon>0$ and therefore
defines a holomorphic function $G = G(u)$ for all $u\in\mathbb{C}$ with Re $u>0$.
Furthermore, there is a constant $M(\varepsilon)>0$ such that

$$|G(u)| \leq M(\varepsilon)|F|_{L_p[0,\infty)} \text{ for all} \qquad\qquad (2.4.14)$$
$u\in\mathbb{C}$ such that Re $u\geq\varepsilon$.

If we put

$$F_n(t) = \sum_{j=1}^{n} a_j^n z_j(t) = \sum_{j=1}^{n} a_j^n e^{-2\pi\lambda_j t}$$

then

$$\lim_{n\to\infty} |F-F_n|_{L_p[0,\infty)} = 0$$

and by the above arguments with respect to $F-F_n (\in A_p(\wedge))$ instead
of F we conclude, for each $\varepsilon > 0$,

$$|G(u) - F_n(u)| \leq M(\varepsilon) \ |F-F_n|_{L_p[0,\infty)}$$

for all $u \in \mathbb{C}$ such that Re $u \geq \varepsilon$.

Therefore $F = F(u)$ has the representation

$$F(u) = \sum_{j=1}^{\infty} a_j e^{-2\pi\lambda_j u} \quad (a_j, \ j \in \mathbb{N}, \ by \ (2.4.5)) \tag{2.4.15}$$

for all $u \in \mathbb{C}$ with Re $u > 0$ and the series (2.4.15) converges
absolutely and uniformly to $F(u)$ for each $u \in \mathbb{C}$ such that Re $u \geq \varepsilon$
where $\varepsilon > 0$ is arbitrary.

Summarizing we can formulate the following

Theorem 2.4.2: Let $(\lambda_j)_{j \in \mathbb{N}}$ be a sequence of positive reals such
that the conditions (2.4.2) and (2.4.9) are satisfied. Then, for
each $F \in A_p(\wedge)$, the corresponding extension to a function $F = F(u)$,
$u \in \mathbb{C}$, is holomorphic for all $u \in \mathbb{C}$ with Re $u > 0$ and has a representation
(2.4.15) such that, for each $\varepsilon > 0$, the series in (2.4.15) converges
absolutely and uniformly to $F(u)$ for every $u \in \mathbb{C}$ with Re $u \geq \varepsilon$.
Furthermore, for each $\varepsilon > 0$, there is a constant $M(\varepsilon) > 0$ such that

$$|F(u)| \leq M(\varepsilon) \ |F|_{L_p[0,\infty)} \quad \text{for all}$$
$$\tag{2.4.16}$$

$u \in \mathbb{C}$ such that Re $u \geq \varepsilon$.

Next we consider exponential sums and series on finite intervals
[0,T], T>0, in order to be able to apply the results to moment
problems as being considered in Sections 2.1.1, 2.1.2 and to be
investigated in Sections 2.4.2, 2.4.3, 2.4.4. The key result is
the following

Theorem 2.4.3: Let the assumptions of Theorem 2.4.2 hold. Then
for each T>0 there is a constant $\gamma_T > 0$ such that

$$|F|_{L_p[0,\infty)} \leq \gamma_T \ |F|_{L_p[0,T]} \quad \text{for all } F \in A_p(\wedge) \tag{2.4.17}$$

Proof: We assume that the assertion is false. Then there exists
a sequence $(F_k)_{k\in\mathbb{N}}$ in $A_p(\wedge)$ with

$$|F_k|_{L_p[0,\infty)} = 1 \text{ for all } k\in\mathbb{N} \tag{2.4.18}$$

and

$$\lim_{k\to\infty} |F_k|_{L_p[0,T]} = 0. \tag{2.4.19}$$

By Theorem 2.4.2 the extension $F_k(u)$, $u\in\mathbb{C}$, of each F_k is a
holomorphic function on $\{u\in\mathbb{C} \mid \mathrm{Re}\ u>0\}$ and for each $\varepsilon>0$ there
is a constant such that

$$|F_k(u)| \le M(\varepsilon)\ |F_k|_{L_p[0,\infty)} = M(\varepsilon)$$

for all $u\in\mathbb{C}$ with $\mathrm{Re}\ u\ge\varepsilon$.

Therefore $(F_k)_{k\in\mathbb{N}}$ is a normal family of holomorphic functions
on $\{u\in\mathbb{C} \mid \mathrm{Re}\ u\ge\varepsilon\}$ and has a subsequence (F_{k_i}) that converges
uniformly to some holomorph function F on $\{u\in\mathbb{C} \mid \mathrm{Re}\ u\ge\varepsilon\}$. If
we choose $\varepsilon\in(0,T)$, then it follows that (F_{k_i}) converges to F
in $L_p[\varepsilon,T]$ and hence

$$\lim_{i\to\infty} |F_{k_i}|_{L_p[\varepsilon,T]} = |F|_{L_p[\varepsilon,T]}.$$

By virtue of (2.4.19) this implies

$$F = 0 \text{ almost everywhere on } [\varepsilon,T].$$

This in turn implies $F \equiv 0$ on $\{u\in\mathbb{C} \mid \mathrm{Re}\ u\ge\varepsilon\}$ since F is holomorphic. From the uniform convergence of (F_{k_i}) to F on $\{u\in\mathbb{C} \mid$
$\mathrm{Re}\ u\ge\varepsilon\}$ it also follows that (F_{k_i}) converges to F in $L_p[\varepsilon,\infty)$,
hence

$$\lim_{i\to\infty} |F_{k_i}|_{L_p[\varepsilon,\infty)} = |F|_{L_p[\varepsilon,\infty)} = 0.$$

Therefore we obtain from

$$|F_k|^p_{L_p[0,\infty)} \le |F_k|^p_{L_p[0,T]} + |F_k|^p_{L_p[\varepsilon,\infty)}$$

that

$$\lim_{i\to\infty} |F_{k_i}|_{L_p[0,\infty)} = 0$$

which contradicts (2.4.18). This completes the proof.

<u>Corollary:</u> Let $(\lambda_j)_{j \in \mathbb{N}}$ be a sequence of positive reals such that the conditions (2.4.2) and (2.4.9) are satisfied. Then the sequence (2.4.1) is minimal in $L_p[0,T]$ for every $T>0$.

<u>Proof:</u> For each $k \in \mathbb{N}$ let $A_p^k(T)$ and $A_p^k(\infty)$ be the closure in $L_p[0,T]$ and $L_p[0,\infty)$, respectively, of the linear space which is generated by the sequence $(z_j)_{j \in \mathbb{N}}$, $j \neq k$. We assume that there is a $k \in \mathbb{N}$ such that $z_k \in A_p^k(T)$. Then there is a sequence $(v^n)_{n \in \mathbb{N}}$ of finite exponential sums formed by z_j's for $j \neq k$ such that

$$\lim_{n \to \infty} |v^n - z_k|_{L_p[0,T]} = 0.$$

Since $z_k \in A_p(\Lambda)$ and $v^n \in A_p(\Lambda)$ for all $n \in \mathbb{N}$, it follows from (2.4.17) that

$$\lim_{n \to \infty} |v^n - z_k|_{L_p[0,\infty)} = 0,$$

hence $z_k \in A_p^k(\infty)$ which contradicts the fact that, by Theorem 2.4.1, the sequence $(z_j)_{j \in \mathbb{N}}$ is minimal in $L_p[0,\infty)$. This completes the proof.

2.4.2. Solvability

As in Section 2.4.1 we consider a strictly increasing sequence $(\lambda_j)_{j \in \mathbb{N}}$ of positive real numbers which satisfies (2.4.2) and (2.4.9). Under these <u>general assumptions</u> for the following we pose the <u>exponential moment problem</u> to be investigated as follows: For a given $T \in (0,\infty]$ and a given sequence $(c_j)_{j \in \mathbb{N}}$ of real numbers which are not all equal to zero find some $u \in L_\infty[0,T]$ such that

$$\int_0^T u(t) e^{-\lambda_j t} dt = c_j \text{ for all } j \in \mathbb{N}. \tag{2.4.20}$$

Obviously this problem is a special case of the abstract moment problem in Section 2.2.1 for $Z = L_1[0,T]$, $X = Z^* = L_\infty[0,T]$ and $z_j(t) = e^{-\lambda_j t}$, $t \in [0,T]$, $j \in \mathbb{N}$. Therefore Theorem 2.2.1 can be applied in order to obtain

<u>Theorem 2.4.4:</u> a) If the moment problem (2.4.20) has a solution $u \in L_\infty[0,T]$, then for every $\gamma \geq |u|_{\infty,T}$ where

$$|u|_{\infty,T} = \underset{t \in [0,T]}{\text{ess sup}} |u(t)|$$

it follows that

$$\sum_{j=1}^{n} c_j y_j \leq \gamma \int_0^T | \sum_{j=1}^{n} y_j e^{-\lambda_j t} | \, dt \qquad (2.4.21)$$

for all $y = (y_1, \ldots, y_n)^T \in \mathbb{R}^n$ and all $n \in \mathbb{N}$.

b) If (2.4.21) is satisfied for some $\gamma > 0$, then there exists a solution $u \in L_\infty[0,T]$ of (2.4.20) with $|u|_{\infty,T} \leq \gamma$.

<u>Remark:</u> For this theorem the general assumptions on the sequence $(\lambda_j)_{j \in \mathbb{N}}$ as made above are not needed. For sequences $(\lambda_j)_{j \in \mathbb{N}}$ which satisfy the general assumptions we can use Theorem 2.4.3 in order to reduce the question of solvability of (2.4.20) for any $T \in (0,\infty]$ to the solvability for $T = \infty$. At first we prove

<u>Theorem 2.4.5:</u> a) For each $T \in (0,\infty]$ there exists a constant $C_T \geq 1$ such that

$$\int_0^\infty | \sum_{j=1}^{n} y_j e^{-\lambda_j t} | \, dt \leq C_T \int_0^T | \sum_{j=1}^{n} y_j e^{-\lambda_j t} | \, dt$$

$$(2.4.22)$$

for all $y = (y_1, \ldots, y_n)^T \in \mathbb{R}^n$ and all $n \in \mathbb{N}$.

b) The constants C_T can be chosen such that

$$C_T \searrow 1 \text{ as } T \to \infty.$$

<u>Proof:</u> Part a) is an immediate consequence of Theorem 2.4.3.

In order to show b) we first observe that $C_1 > 1$ for all choices of C_T and that the C_T's can be choosen such that $C_{T_2} \leq C_{T_1}$ for $0 < T_1 < T_2 \leq \infty$.

Therefore it suffices to show that

$$\lim_{n \to \infty} C_n = 1$$

Now, for every $y = (y_1,\dots,y_n)^T \in \mathbb{R}^n$ and $n \in \mathbb{N}$,

$$\int_1^\infty |\sum_{j=1}^n y_j e^{-\lambda_j t}| \, dt = \int_0^\infty |\sum_{j=1}^n y_j e^{-\lambda_j t}| \, dt - $$

$$\underbrace{\int_0^1 |\sum_{j=1}^n y_j e^{-\lambda_j t}| \, dt}_{\geq \frac{1}{c_1} \int_0^\infty \dots \, dt} \leq \rho \int_0^\infty |\sum_{j=1}^n y_j e^{-\lambda_j t}| \, dt$$

where $\rho = 1 - \frac{1}{c_1} \in (0,1)$.

Obviously the identity

$$\int_{k+1}^\infty |\sum_{j=1}^n y_j e^{-\lambda_j t}| \, dt = \int_k^\infty |\sum_{j=1}^n (y_j e^{-\lambda_j}) e^{-\lambda_j t}| \, dt$$

holds for all $k \in \mathbb{N}$. Further the inequality

$$\int_k^\infty |\sum_{j=1}^n y_j e^{-\lambda_j t}| \, dt \leq \rho^k \int_0^\infty |\sum_{j=1}^n y_j e^{-\lambda_j t}| \, dt$$

is true for $k = 1$ and for all $y = (y_1,\dots,y_n)^T \in \mathbb{R}^n$ and all $n \in \mathbb{N}$ as seen above. We assume this to hold for some $k \in \mathbb{N}$. Then

$$\int_{k+1}^\infty |\sum_{j=1}^n y_j e^{-\lambda_j t}| \, dt = \int_k^\infty |\sum_{j=1}^n (y_j e^{-\lambda_j}) e^{-\lambda_j t}| \, dt$$

$$\leq \rho^k \int_0^\infty |\sum_{j=1}^n (y_j e^{-\lambda_j}) e^{-\lambda_j t}| \, dt$$

$$= \rho^k \int_1^\infty |\sum_{j=1}^n y_j e^{-\lambda_j t}| \, dt \leq \rho^{k+1} \int_0^\infty |\sum_{j=1}^n y_j e^{-\lambda_j t}| \, dt.$$

For every $k \in \mathbb{N}$, every $y = (y_1,\dots,y_n)^T \in \mathbb{R}^n$ and $n \in \mathbb{N}$ we therefore conclude

$$\int_0^\infty |\sum_{j=1}^n y_j e^{-\lambda_j t}| \, dt = \int_0^k |\sum_{j=1}^n y_j e^{-\lambda_j t}| \, dt + \int_k^\infty |\sum_{j=1}^n y_j e^{-\lambda_j t}| \, dt$$

$$\leq \int_0^k |\sum_{j=1}^n y_j e^{-\lambda_j t}| \, dt + \rho^k \int_0^\infty |\sum_{j=1}^n y_j e^{-\lambda_j t}| \, dt$$

which implies

$$\int_0^\infty | \sum_{j=1}^n y_j e^{-\lambda_j t} | \, dt \le C_k \int_0^k | \sum_{j=1}^n y_j e^{-\lambda_j t} | \, dt$$

with $C_k = (1-\rho^k)^{-1} \searrow 1$ as $k \to \infty$. This completes the proof of part b).

For each sequence $\lambda = (\lambda_j)_{j\in\mathbb{N}}$, each sequence $c = (c_j)_{j\in\mathbb{N}}$ and each $T\in(0,\infty]$ we define

$$S(T,\lambda,c) = \{u\in L_\infty[0,T] \mid u \text{ satisfies } (2.4.20)\}. \qquad (2.4.23)$$

From Theorems 2.4.4 and 2.4.5 we deduce the

Theorem 2.4.6: a) If $S(\infty,\lambda,c)$ is nonempty, then $S(T,\lambda,c)$ is nonempty for every $T\in(0,\infty]$.

b) If $u\in S(\infty,\lambda,c)$, then for each $\gamma>|u|_{\infty,\infty}$ there exist $T\in(0,\infty)$ and some $\tilde{u}\in S(T,\lambda,c)$ with $|\tilde{u}|_{\infty,T}<\gamma$.

Proof: a) Let $u\in S(\infty,\lambda,c)$ be chosen. Then by Theorem 2.4.4a) for $T = \infty$ it follows that

$$\sum_{j=1}^n c_j y_j \le |u|_{\infty,\infty} \int_0^\infty | \sum_{j=1}^n y_j e^{-\lambda_j t} | \, dt$$

for all $y = (y_1,\ldots,y_n)^T\in\mathbb{R}^n$ and all $n\in\mathbb{N}$. For each $T\in(0,\infty]$ we then obtain, by Theorem 2.4.5, the existence of a constant $C_T\ge 1$ such that

$$\sum_{j=1}^n c_j y_j \le |u|_{\infty,\infty} \cdot C_T \int_0^T | \sum_{j=1}^n y_j e^{-\lambda_j t} | \, dt$$

for all $y = (y_1,\ldots,y_n)^T\in\mathbb{R}^n$ and all $n\in\mathbb{N}$. Again, by Theorem 2.4.4b), this implies the existence of some $\tilde{u}\in S(T,\lambda,c)$ with $|\tilde{u}|_{\infty,T} \le |u|_{\infty,\infty} \cdot C_T$. This, in connection with Theorem 2.4.5b), also implies b).

Part a) of Theorem 2.4.6 can be strengthened to the

Theorem 2.4.7: If there exists a measurable function $g : [0,\infty]\to\mathbb{R}$ and $\varepsilon\in[0,\lambda_1)$ such that $f \cdot g\in L_\infty[0,\infty]$ where $f(t) = e^{-\varepsilon t}$, $t\in[0,\infty]$ and

$$\int_O^\infty g(t)e^{-\lambda_j t} \, dt = c_j \text{ for all } j\in\mathbb{N},$$

then $S(T,\lambda,c)$ is nonempty for every $T\in(0,\infty]$.

Proof: Let $n\in\mathbb{N}$ and $y = (y_1,\dots,y_n)^T\in\mathbb{R}^n$ be given together with some $T\in(0,\infty]$. Then

$$\sum_{j=1}^n c_j y_j \le \int_O^\infty \left| \sum_{j=1}^n y_j e^{-(\lambda_j-\epsilon)t} (e^{-\epsilon t} g(t)) \right| dt$$

$$\le \| f\cdot g\|_{\infty,\infty} \int_O^\infty \left| \sum_{j=1}^n y_j e^{-\mu_j t} \right| dt$$

where $(\mu_j = \lambda_j - \epsilon)_{j\in\mathbb{N}}$ is an increasing sequence of positive real numbers with $\sum_{j=1}^\infty \dfrac{1}{\mu_j} < +\infty$. Therefore, by Theorem 2.4.5a), there exists some $C_T \ge 1$ such

$$\sum_{j=1}^n c_j y_j \le \| f\cdot g\|_{\infty,\infty} C_T \int_O^T \left| \sum_{j=1}^n y_j e^{-\mu_j t} \right| dt.$$

By Theorem 2.4.4b) there exists some $u\in L_\infty[0,T]$ with

$$\int_O^T u(t)e^{-\mu_j t} \, dt = c_j \text{ for all } j\in\mathbb{N}$$

which implies

$$\int_O^T v(t)e^{-\lambda_j t} \, dt = c_j \text{ for all } j\in\mathbb{N}$$

where $v = (^u/f)\in L_\infty[0,T]$. This completes the proof.

Corollary: $S(T,\lambda,c)$ is nonempty for every $T\in(0,\infty]$, if the sequence $c = (c_j)_{j\in\mathbb{N}}$ is given by

$$c_j = \sum_{k=1}^N a_k \lambda_j^{-k}$$

for some $N\in\mathbb{N}$ and $(a_1,\dots,a_N)^T\in\mathbb{R}^N$.

Proof: Because of

$$\lambda_j^{-k} = \frac{1}{(k-1)!} \int_0^\infty e^{-\lambda_j t} t^{k-1} \, dt, \quad j \in \mathbb{N},$$

the function g in Theorem 2.4.7 can be chosen as

$$g(t) = \sum_{k=1}^{N} \frac{a_k}{(k-1)!} t^{k-1}, \quad t \in [0,\infty).$$

A case of particular interest is the case where N = 1 and

$$c_j = \frac{a_1}{\lambda_j} \quad \text{for all } j \in \mathbb{N} \text{ and some } a_1 \in \mathbb{R}.$$

Here we can also apply Theorem 2.4.6 in order to see that $S(T,\lambda,c)$ is nonempty for every $T \in (0,\infty]$. Since for $u(t) = a_1$ for all $t \in [0,\infty)$ we see that

$$\int_0^\infty u(t) e^{-\lambda_j t} \, dt = \frac{a_1}{\lambda_j} = c_j \quad \text{for all } j \in \mathbb{N}$$

so that part a) is applicable. In addition, by part b), for every $\gamma > |a_1|$ there is a $T \in (0,\infty]$ and some $\tilde{u} \in S(T,\lambda,c)$ with $|\tilde{u}|_{\infty,T} < \gamma$.

There exists another important class of sequences $c = (c_j)_{j \in \mathbb{N}}$ for which $S(\infty,\lambda,c)$ is nonempty and hence, by Theorem 2.4.6, $S(T,\lambda,c)$ is nonempty for every $T \in (0,\infty]$. In order to determine this class we first observe that from Theorem 2.4.1 and the proof of Theorem 2.4.2, for each choice of $\alpha > 0$, there exist a sequence $(x_k)_{k \in \mathbb{N}}$ in $L_\infty[0,\infty]$ and a constant $K(\alpha) > 0$ such that

$$\int_0^\infty x_k(t) e^{-\lambda_j t} \, dt = \delta_{kj} = \begin{cases} 0, & \text{if } k \neq j, \\ 1, & \text{if } k = 1, \end{cases} \qquad (2.4.24)$$

for all $k, j \in \mathbb{N}$ and

$$|x_k|_{\infty,\infty} \le K(\alpha) e^{\alpha \lambda_k} \quad \text{for all } k \in \mathbb{N} \qquad (2.4.25)$$

on condition that there is a constant $\delta > 0$ such that

$$\lambda_{j+1} - \lambda_j \ge \delta \quad \text{for all } j \in \mathbb{N}$$

$$\text{and } \lambda_1 \ge \delta. \qquad (2.4.9)$$

On using the sequence $(x_k)_{k \in \mathbb{N}}$ a formal solution of (2.4.20) for $T = \infty$ can be written down as

$$u = \sum_{k=1}^{\infty} c_k x_k .$$

The question, however, is whether this series converges in $L_{\infty}[0,\infty]$. This is guaranteed by the assumptions of

__Theorem 2.4.8:__ Let the sequence $(\lambda_j)_{j \in \mathbb{N}}$ satisfy (2.4.2) and (2.4.9). Then for each sequence $(c_j)_{j \in \mathbb{N}} \in \ell_1$ such that there is some $\delta > 0$ with

$$\sum_{j=1}^{\infty} |c_j| \, e^{\delta \lambda_j} < \infty \tag{2.4.26}$$

the set $S(\infty, \lambda, c)$ is nonempty.

__Proof:__ Let $(x_k)_{k \in \mathbb{N}}$ be any sequence in $L_{\infty}[0,\infty]$ with (2.4.24) and (2.4.25) for $\alpha = \delta$. For each pair n_1, $n_2 \in \mathbb{N}$ with $n_1 < n_2$ we then conclude

$$| \sum_{k=n_1}^{n_2} c_k x_k |_{\infty, \infty} \leq \sum_{k=n_1}^{n_2} |c_k| \, |x_k|_{\infty, \infty} \leq K(\delta) \sum_{k=n_1}^{n_2} |c_k| \, e^{\delta \lambda_k}$$

so that, by virtue of (2.4.26) the sequence $(\sum_{k=1}^{n} c_k x_k)_{n \in \mathbb{N}}$ is a Cauchy sequence in $L_{\infty}[0,\infty]$ which converges to

$$u = \sum_{k=1}^{\infty} c_k x_k \in S(\infty, \lambda, c) .$$

This completes the proof.

In the case of exponential moment problems which arise with problems of heat control, as in Section 2.1, a stronger condition than (2.4.2) is satisfied, namely, the existence of constants $\alpha \in \mathbb{R}$ and $K > 0$ such that

$$\lambda_k = K(k+\alpha)^2 + o(k) \text{ as } k \to \infty . \tag{2.4.27}$$

Under this condition (which implies (2.4.2)) the assertion of Theorem 2.4.8 can be made without assuming (2.4.9) and under a weaker condition than (2.4.26).

In order to show this we first summarize the statement of
Theorem 2.4.1 and the following remarks to the assertion
that there exists a sequence $(x_k)_{k \in \mathbb{N}}$ in $L_\infty(0,\infty)$ with (2.4.24)
and some constant $C > 0$ such that

$$\|x_k\|_{\infty,\infty} \leq C\lambda_k(1+\lambda_k)^2 \prod_{\substack{j=1 \\ j \neq k}}^{\infty} \frac{\lambda_j + \lambda_k}{|\lambda_j - \lambda_k|}. \qquad (2.4.28)$$

If, in addition, there exist constants $a \in \mathbb{R}$ and $K > 0$ with
(2.4.27), then, by Fattorini/Russell [3],

$$\prod_{\substack{j=1 \\ j \neq k}}^{\infty} \frac{\lambda_j + \lambda_k}{|\lambda_j - \lambda_k|} = \exp\left[(\frac{\pi}{\sqrt{K}} + \zeta_k)\lambda_k^{1/2}\right] \quad \text{as } k \to \infty$$

with $\lim_{k \to \infty} \zeta_k = 0$. In connection with (2.4.28) this implies the
existence of a constant $D > 0$ such that

$$\|x_k\|_{\infty,\infty} \leq D\lambda_k(1+\zeta_k)^2\exp\left[(\frac{\pi}{\sqrt{K}} + \zeta_k)\lambda_k^{1/2}\right]$$

$$\text{for all } k \in \mathbb{N} \qquad (2.4.29)$$

and $\lim_{k \to \infty} \zeta_k = 0$.

Instead of Theorem 2.4.8 we then have

Theorem 2.4.9: Let $(\lambda_j)_{j \in \mathbb{N}}$ be a strictly increasing sequence of
positive real numbers such that (2.4.27) is satisfied for some
$a \in \mathbb{R}$ and $K > 0$. Then, for each sequence $(c_j)_{j \in \mathbb{N}} \in \ell_1$ such that

$$\sum_{j=1}^{\infty} |c_j| \exp\left[(\frac{\pi}{\sqrt{K}} + \zeta)\lambda_j^{1/2}\right] < \infty \qquad (2.4.30)$$

for some $\zeta > 0$, the set $S_\infty(\infty,\lambda,c)$ is nonempty. The proof is
similar to the one of Theorem 2.4.8 and makes use of the fact
that, by virtue of (2.4.29), for every $\zeta > 0$ there is some constant
$D(\zeta) > 0$ such that

$$\|x_k\|_{\infty,\infty} \leq D(\zeta) \exp\left[(\frac{\pi}{\sqrt{K}} + \zeta)\lambda_k^{1/2}\right] \text{ for all } k \in \mathbb{N}.$$

2.4.3. On Least Norm Solutions.

We define, for every $T \in (0, \infty]$,

$$\gamma(T, \lambda, c) = \inf\{ |u|_{\infty, T} \mid u \in S(T, \lambda, c) \} \tag{2.4.31}$$

with $S(T, \lambda, c)$ being given by (2.4.23).

Without the general assumptions at the beginning of Section 2.4.2 we have, as a consequence of Theorem 2.2.3, the

Theorem 2.4.10: If $S(T, \lambda, c)$ is nonempty for some $T \in (0, \infty]$ and some sequence $c = (c_j)_{j \in \mathbb{N}}$, then there exists a least norm solution $\hat{u} \in L_\infty[0, T]$ of (2.4.20), i.e., a solution which satisfies

$$|\hat{u}|_{\infty, T} = \gamma(T, \lambda, c).$$

If (2.4.2) is satisfied, then, for every $T \in (0, \infty]$, by the definition

$$S_T(u) = \sum_{j=1}^{\infty} \int_0^T u(t) e^{-\lambda_j t} dt \, e_j \tag{2.4.32}$$

for $u \in L_\infty[0, T]$ and e_j, $j \in \mathbb{N}$, being the unit vector in ℓ^1 having 1 as j-th component and 0 elsewhere we obtain a continuous linear mapping from $L_\infty[0, T]$ into ℓ^1, since for $z_j(t) = e^{-\lambda_j t}$, $t \in [0, T]$, $j \in \mathbb{N}$, it follows that

$$\sum_{j=1}^{\infty} |z_j|_{1, T} \leq \sum_{j=1}^{\infty} \frac{1}{\lambda_j} < \infty$$

where

$$|z|_{1, T} = \int_0^T |z(t)| \, dt \text{ for } z \in Z = L_1[0, T]$$

(see Sections 2.2.2 and 2.3.1).

Obviously (2.4.20) is equivalent to $S_T(u) = c$. Therefore $S(T, \lambda, c)$ is nonempty, if and only if there exists some $u \in L_\infty[0, T]$ with $S_T(u) = c$.

Moreover, the nuclear operator S_T (2.4.32) is compact as a limit of compact operators

$$S_T^N(u) = \sum_{j=1}^{N} \int_0^T u(t) e^{-\lambda_j t} dt \, e_j, \quad u \in L_\infty[0, T], \ N \in \mathbb{N},$$

and therefore maps weak[*] convergent into norm-convergent
sequences (see Section 2.4.4.1). Hence Theorem 2.4.10 can
also be derived from Theorem 1.5.2, if the condition (2.4.2)
is satisfied.

For the following we assume that $\lambda = (\lambda_j)_{j \in \mathbb{N}}$ satisfies (2.4.2)
and (2.4.9). Our aim is to derive a bang-bang principle for
least norm solutions of (2.4.20) on using Theorem 1.5.4 (see
also Theorem 2.2.5). To this end let

$$R_T = S_T(L_\infty[0,T]). \tag{2.4.33}$$

As in Section 2.2.2 we define a norm in R_T by

$$|c|_{R_T} = \inf\{|u|_{\infty,T} \mid S_T(u) = c\} \tag{2.4.34}$$

for every $c \in R_T$. With this norm R_T becomes a Banach space and
$S_T : L_\infty[0,T] \to R_T$ turns out to be a continuous linear mapping with
$|S_T| \le 1$ (see Section 2.2.2).

Let $\varepsilon > 0$ be chosen arbitrarily. Then we define a linear mapping
$V_\varepsilon : \ell^1 \to \ell^1$ by

$$V_\varepsilon c = (c_j e^{-\lambda_j \varepsilon})_{j \in \mathbb{N}}, \quad c = (c_j)_{j \in \mathbb{N}} \in \ell^1.$$

Obviously V_ε is continuous, since $|V_\varepsilon c|_{\ell^1} \le |c|_{\ell^1}$ for all $c \in \ell^1$.
By Theorem 2.4.8 it follows that

$$V_\varepsilon(\ell^1) \subset R_T \text{ for all } \varepsilon > 0,$$

since for $\varepsilon > 0$ and $c \in V_\varepsilon(\ell^1)$ being given it follows that

$$\sum_{j=1}^\infty e^{\lambda_j \varepsilon} |c_j| < \infty.$$

Therefore $S(\infty, \lambda, c)$ and, by Theorem 2.4.6a), $S(T, \lambda, c)$ is nonempty
which is equivalent to saying that $c \in R_T$.

Moreover, $V_\varepsilon : \ell^1 \to R_T$ is continuous by the following argument:
Let $(c^k, V_\varepsilon c^k)_{k \in \mathbb{N}}$ in $\ell^1 \times R_T$ be a sequence in the graph of V_ε
such that

$$|c^k - c|_{\ell^1} \to 0 \text{ and } |V_\varepsilon c^k - y|_{R_T} \to 0 \text{ as } k \to \infty$$

for some $c \in \ell^1$ and some $y \in R_T$. Because of

$$|V_\varepsilon c^k - y|_{\ell^1} \leq (\sum_{j=1}^{\infty} \frac{1}{\lambda_j}) |V_\varepsilon c^k - y|_{R_T} \quad \text{for all } k \in \mathbb{N}$$

it follows that

$$|V_\varepsilon c^k - y|_{\ell^1} \to 0 \text{ as } k \to \infty \text{ and } y = V_\varepsilon c$$

due to the continuity of $V_\varepsilon : \ell^1 \to \ell^1$. Therefore the graph of V_ε is closed in $\ell^1 \times R_T$ and, by the closed graph theorem, $V_\varepsilon : \ell^1 \to R_T$ is continuous.

Obviously we have

$$V_{\varepsilon_1} (\ell^1) \subset V_{\varepsilon_2} (\ell^1), \quad \text{if } 0 < \varepsilon_2 \leq \varepsilon_1.$$

Therefore $V = \bigcup_{\varepsilon > 0} V_\varepsilon (\ell^1)$ is a linear subspace of R_T. Let W be the closure of V in R_T and let X be the counterimage $S_T^{-1}(W)$. Then X is a closed subspace of $L_\infty[0,T]$ and hence a Banach space as well as W being a closed subspace of the Banach space R_T. By constuction, $S_T(X) = W$.

Therefore, by Theorem 1.5.4, for every $c \in W$ with $c \neq \theta_{\ell^1}$ there exists a $y^* \in W^*$, $y^* \neq \theta_{W^*}$, such that

$$y^* (S_T \hat{u}) = |S_T^*(y^*)| \, |\hat{u}|_{\infty, T} \tag{2.4.35}$$

for every $\hat{u} \in S(T, \lambda, c)$ with $|\hat{u}|_{\infty, T} = \gamma(T, \lambda, c)$ (see (2.4.31)) where $S_T^* : W^* \to X^*$ is the adjoint operator os $S_T : X \to W$.

In order to exploit (2.4.35) is is necessary to find a representation of $y^* (S_T \hat{u})$. For this purpose we define, for every $\varepsilon \in (0, T)$ and every $u \in L_\infty[0,T]$ the operator $S_{T,\varepsilon} : L_\infty[0,T] \to \ell^1$ by

$$S_{T,\varepsilon} u = \sum_{j=1}^{\infty} \int_\varepsilon^T e^{-\lambda_j(t-\varepsilon)} u(t) \, dt \, e_j$$

where again e_j, $j \in \mathbb{N}$, denotes the j-th unit vector in ℓ^1. Then, for every $u \in L_\infty[0,T]$ with $u = 0$ a.e. on $(0, \varepsilon)$ it follows that

$$S_T u = V_\varepsilon S_{T,\varepsilon} u,$$

consequently,

$$y^*(S_T u) = V_\varepsilon^*(y^*)(S_{T,\varepsilon} u)$$

where $V_\varepsilon^* : W^* \to \ell^\infty$ is the adjoint operator of $V_\varepsilon : \ell^1 \to W$. Now let $c \in \ell^1$ be chosen arbitrarily and let $d = V_\varepsilon c$. Then we define, for each $N \in \mathbb{N}$,

$$d_N = (0, \ldots, 0, d_{N+1}, d_{N+2}, \ldots)$$

and put

$$u_N = \sum_{j=N+1}^{\infty} d_j x^j$$

where $(x^j)_{j \in \mathbb{N}}$ is a sequence in $L_\infty[0, \infty]$ with (2.4.24) and (2.4.25) for $\alpha = \varepsilon$. Then

$$|u_N|_{\infty, \infty} \leq \sum_{j=N+1}^{\infty} |d_j| \; |x^j|_{\infty, \infty} \leq K(\varepsilon) \sum_{j=N+1}^{\infty} |d_j| \; e^{\varepsilon \lambda_j} \to 0$$

as $N \to \infty$. And since $S_\infty u_N = d_N$ for all $N \in \mathbb{N}$, it follows that

$$\lim_{N \to \infty} |d_N|_{R_\infty} = O \qquad \lim_{N \to \infty} |d_N|_{R_T} = O$$

due to the equivalence of all norms $|\cdot|_{R_T}$ for $T \in (0, \infty]$ which will be proved later.

If we put $d^N = d - d_N = (d_1, \ldots, d_N, 0, \ldots, 0)$, then we see that

$$\lim_{N \to \infty} |d^N - d|_{R_T} = \lim_{N \to \infty} |d^N - V_\varepsilon c|_{R_T} = O$$

and therefore

$$y^*(V_\varepsilon c) = \lim_{N \to \infty} y^*(d^N) = \lim_{N \to \infty} \sum_{j=1}^{N} d_j \; y^*(e_j)$$

$$= \sum_{j=1}^{\infty} e^{-\lambda_j \varepsilon} y^*(e_j) c_j = V_\varepsilon^*(y^*)(c)$$

with

$$V_\varepsilon^*(y^*) = (e^{-\lambda_j \varepsilon} y^*(e_j))_{j \in \mathbb{N}} \in \ell^\infty.$$

This result in turn leads to

$$y^*(S_T u) = v^*_\varepsilon(y^*)(S_{T,\varepsilon}u)$$

$$= \sum_{j=1}^{\infty} e^{-\lambda_j \varepsilon} y^*(e_j) \int_{\varepsilon}^{T} e^{-\lambda_j(t-\varepsilon)} u(t) \, dt$$

$$= \sum_{j=1}^{\infty} y^*(e_j) \int_{\varepsilon}^{T} e^{-\lambda_j t} u(t) \, dt$$

Let N and M∈N be given. Then

$$\sum_{j=N}^{N+M} |y^*(e_j)| \int_{\varepsilon}^{T} e^{-\lambda_j t}$$

$$\leq \sum_{j=N}^{N+M} |y^*(e_j)| e^{-\lambda_j \varepsilon} \frac{1}{\lambda_j} (1-e^{-\lambda_j(T-\varepsilon)})$$

$$\leq \sum_{j=N}^{N+M} \frac{1}{\lambda_j} \sup_{j \in N} |y^*(e_j)| e^{-\lambda_j \varepsilon} \to 0$$

as N → ∞. Therefore $(\sum_{j=1}^{N} y^*(e_j)e^{-\lambda_j t})_{N \in N}$, t∈[ε,T], is a Cauchy

sequence in $L_1[\varepsilon,T]$ and hence converges to

$$\sum_{j=1}^{\infty} y^*(e_j)e^{-\lambda_j \cdot} \in L_1[\varepsilon,T].$$

As a result we conclude

$$y^*(S_T u) = \int_{\varepsilon}^{T} w(t) \, u(t) \, dt$$

for every ε∈(0,T) and every u∈$L_\infty[0,T]$ with u = 0 a.e. on (0,ε) where

$$w = \sum_{j=1}^{\infty} y^*(e_j)e^{-\lambda_j \cdot} \quad L_1[\varepsilon,T]$$

is independent of ε.

Let us assume that w = 0 a.e. on (ε,T) for all ε∈(0,T). Since, by the Corollary to the Theorem 2.4.3, the sequence $(e^{-\lambda_j t})_{j \in N}$, t∈[0,T] is minimal on [0,T], it follows that

$$y^*(e_j) = 0 \text{ for all } j \in N.$$

Let $c \in \ell^1$ be given. Then we define $c^N = (c_1, \ldots, c_N, 0, 0, \ldots)$ for each $N \in \mathbb{N}$ so that $\lim\limits_{N \to \infty} |c^N - c|_{\ell^1} = 0$. Since

$$V_\varepsilon(c^N) = \sum_{j=1}^{N} e^{-\lambda_j \varepsilon} c_j e_j \text{ for all } N \in \mathbb{N} \text{ and all } \varepsilon > 0,$$

it follows that

$$y^*(V_\varepsilon c^N) = 0 \text{ for all } N \in \mathbb{N} \text{ and all } \varepsilon > 0,$$

consequently, $y^*(V_\varepsilon c) = 0$ because $V_\varepsilon : \ell_1 \to R_T$ is continuous. As a result we have $y^* = 0$ on V and therefore also on W which contradicts the above choice of y^*. Therefore the assumption $w = 0$ a.e. on (ε, T) for all $\varepsilon \in (0, T)$ is false and there exists an $\varepsilon \in (0, T)$ such that $w \neq 0$ on a subset I of (ε, T) of positive measure. By construction w belongs to the closure $A_1(\wedge)$ of the linear subspace which is generated by the sequence (2.4.1) in $L_1[0, T]$ and hence, by Theorem 2.4.3, in $L_1[0, \infty)$. By Theorem 2.4.2 the function w is real-analytic on (ε, ∞) for all $\varepsilon > 0$ which implies

$$w \neq 0 \text{ on } (\varepsilon, T) \text{ except in finitely many points,} \qquad (2.4.36)$$
$$\text{for every } \varepsilon \in (0, T).$$

Summarizing we have the

__Theorem 2.4.11:__ For every $y^* \in W^*$ with $y^* \neq \Theta_{W^*}$ and every $\varepsilon \in (0, T)$ there exists a function $w \in L_1[\varepsilon, T]$ with (2.4.36) which is independent of ε such that

$$y^*(S_T u) = \int_\varepsilon^T w(t) u(t)\, dt$$

for all $u \in L_\infty[0, T]$ with $u = 0$ a.e. on $(0, \varepsilon)$.

After these intermediate considerations we return to the minimum norm problem. First we deduce from (2.4.35) that

$$y^*(S_T \hat{u}) \geq |S_T^*(y^*)| \; |u|_{\infty, T} \geq y^*(S_T u) \qquad (2.4.37)$$
$$\text{for every } u \in X \text{ with } |u|_{\infty, T} \leq |\hat{u}|_{\infty, T}.$$

Now let $u \in L_\infty[0, T]$ with $|u|_{\infty, T} \leq |\hat{u}|_{\infty, T}$ be given arbitrarily. For each $\varepsilon \in (0, T)$ we then define

$$u_\varepsilon = (1-\chi_\varepsilon)\, \hat{u} + \chi_\varepsilon u$$

where

$$\chi_\varepsilon(t) = \begin{cases} 0 \text{ for } t \in (0,\varepsilon), \\ 1 \text{ for } t \in [\varepsilon,T]. \end{cases}$$

Then we obtain some $u_\varepsilon \in L_\infty[0,T]$ with

$$|u_\varepsilon|_{\infty,T} \leq |\hat{u}|_{\infty,T} \text{ and } \hat{u} - u_\varepsilon = \chi_\varepsilon(\hat{u}-u).$$

Moreover, we conclude for

$$d_j = \int_0^T u_\varepsilon(t) e^{-\lambda_j t}\, dt, \quad j \in \mathbb{N},$$

that

$$|d_j| \leq |u_\varepsilon|_{\infty,T}\, e^{-\lambda_j \varepsilon}\, \frac{1}{\lambda_j}\, (1-e^{-\lambda_j(T-\varepsilon)}) \leq |u_\varepsilon|_{\infty,T}\, e^{-\lambda_j \varepsilon}\, \frac{1}{\lambda_j}$$

for all $j \in \mathbb{N}$, hence

$$\sum_{j=1}^\infty |d_j|\, e^{\lambda_j} < \infty,$$

which implies $d = V_\varepsilon \tilde{d}$ for some $\tilde{d} \in \ell_1$ and therefore $u_\varepsilon \in X$. As a result we conclude from (2.4.37) and Theorem 2.4.11

$$0 \leq y^*(S_T(\hat{u}-u_\varepsilon)) = \int_\varepsilon^T w(t)(\hat{u}(t)-u(t))\, dt \qquad (2.4.38)$$

for every $\varepsilon \in (0,T)$ and all $u \in L_\infty[0,T]$ with $|u|_{\infty,T} \leq |\hat{u}|_{\infty,T}$ where $w \in L_1[\varepsilon,T]$ is independent of ε and satisfies (2.4.36). This leads to the above mentioned bang-bang-principle for least norm solutions.

__Theorem 2.4.12:__ For every $c \in W$, $c \neq \theta_{\ell_1}$, and every $T \in (0,\infty]$ there exists a function $w \in L_1[\varepsilon,T]$ for all $\varepsilon \in (0,T)$ which is independent of ε and satisfies (2.4.36) such that for every $\hat{u} \in S(T,\lambda,c)$ with $|\hat{u}|_{\infty,T} = \gamma(T,\lambda,c)$ (see (2.4.31))

$$\hat{u}(t) = |\hat{u}|_{\infty,T} \text{ sgn } w(t) \text{ for almost all } t \in [0,T] \qquad (2.4.39)$$

and hence $\hat{u}$ is unique.

<u>Proof</u>: According to the above considerations there exists a function $\dot{w}$ with all the properties as stated in the theorem such that (2.4.38) holds for all $\varepsilon \in (0,T)$ and all $u \in L_\infty[0,T]$ with $|u|_{\infty,T} \leq |\hat{u}|_{\infty,T}$.

Let us assume (2.4.39) to be violated. Then there is a subset I of $[0,T]$ of positive measure such that

$$\hat{u}(t)\omega(t) < |\hat{u}|_{\infty,T}|\omega(t)| \text{ for all } t \in I.$$

We can assume the existence of some $\varepsilon \in (0,T)$ such that $I_\varepsilon = [\varepsilon,T] \cap I$ is of positive measure. If we define

$$u^*(t) = \begin{cases} \hat{u}(t) \text{ for all } t \notin I_\varepsilon, \\ \\ \|\hat{u}\|_{\infty,T} \text{ sgn } w(t) \text{ for all } t \in I_\varepsilon, \end{cases}$$

then $u^* \in L_\infty[0,T]$, $|u^*|_{\infty,T} = |u|_{\infty,T}$ and

$$\int_\varepsilon^T w(t)u^*(t) \, dt = \int_{I_\varepsilon} |\hat{u}|_{\infty,T} |w(t)| + \int_{[\varepsilon,T]\backslash I_\varepsilon} w(t)\hat{u}(t) \, dt$$
$$> \int_\varepsilon^T w(t)\hat{u}(t) \, dt$$

which contradicts (2.4.38) for all $\varepsilon \in (0,T)$ and all $u \in L_\infty[0,T]$ with $|u|_{\infty,T} \leq |\hat{u}|_{\infty,T}$. This contradiction shows that (2.4.39) is true which completes the proof.

<u>Remark</u>: From (2.4.36) it follows that the unique $\hat{u} \in S(T,\lambda,c)$ with $|\hat{u}|_{\infty,T} = \gamma(T,\lambda,c)$ has at most a finite number of zeros on each interval $[\varepsilon,T]$ with $\varepsilon \in (0,T)$.

2.4.4. On Time-Minimal Solutions.
2.4.4.1. Reduction to Least Norm Solutions.

We again assume the conditions (2.4.2) and (2.4.9) to be satisfied so that by (2.4.32), for every $T \in [0,\infty]$, a continuous linear mapping $S_T : L_\infty[0,\infty] \to \ell^1$ is defined. We put $X = L_\infty[0,\infty]$ and

$$|u|_\infty = |u|_{\infty,\infty} \text{ for all } u \in X.$$

Obviously, $S_0(X) = \{\Theta_{\ell_1}\}$ and, for each $\hat{T} \in [0, \infty)$, one can prove that

$$\lim_{T \to \hat{T}+0} |S_T - S_{\hat{T}}| = 0.$$

For each $T \in [0, \infty]$ and each $N \in \mathbb{N}$ we define

$$S_T^N(u)_j = \begin{cases} \int_0^T u(t) e^{-\lambda_j t} \, dt & \text{for } j = 1, \ldots, N, \\ 0 & \text{for } j > N. \end{cases}$$

Then $S_T^N : X \to \ell^1$ is a continuous linear mapping with a finite-dimensional range and therefore maps weak* convergent sequences into norm-convergent sequences. Moreover, we have, for each $u \in X$,

$$|S_T(u) - S_T^N(u)|_{\ell^1} \leq \sum_{j=N+1}^{\infty} \frac{1}{\lambda_j} |u|_{\infty}$$

and therefore

$$\lim_{N \to \infty} |S_T(u) - S_T^N(u)|_{\ell_1} = 0.$$

Let $(u_k)_{k \in \mathbb{N}}$ in X and $u \in X$ be given such that $u_k \overset{*}{\to} u$. Then for every k, $N \in \mathbb{N}$ we have

$$|S_T(u) - S_T(u_k)|_{\ell^1} \leq |S_T(u) - S_T^N(u)|_{\ell^1}$$

$$+ |S_T^N(u) - S_T^N(u_k)|_{\ell^1}.$$

Let $\varepsilon > 0$ be given, Then we chose N so large that $|S_T(u) - S_T^N(u)|_{\ell^1} \leq \frac{\varepsilon}{2}$ which is always possible. Furthermore,

$$|S_T^N(u) - S_T^N(u_k)|_{\ell^1} \leq \frac{\varepsilon}{2} \quad \text{for all } k \geq k(\varepsilon),$$

since S_T^N maps weak* convergent sequences into norm-convergent sequences. As a result we conclude

$$|S_T(u) - S_T(u_k)|_{\ell^1} \leq \varepsilon \quad \text{for all } k \geq k(\varepsilon),$$

i.e., S_T also maps weak* convergent sequences into norm-convergent sequences.

Let $M > 0$ be given such that, for some $c \in \ell^1$, $c \neq \Theta_{\ell^1}$, the set $S(\infty, \lambda, c)$ (see (2.4.23)) is nonempty and $M > \gamma(\infty, \lambda, c)$ (see (2.4.31)). Then the minimum time

$$T(M) = \inf\{T \in [0, \infty] \mid S_T(u) = c \text{ for some}$$

$$u \in X \text{ with } |u|_\infty \leq M\} \tag{2.4.40}$$

is well defined by Theorem 2.4.6b) and from Theorem 1.5.1 we conclude that $T(M) > 0$ and there exists some $u_M \in X$ with

$$S_{T(M)} u_M = c \text{ and } |u_M|_\infty \leq M. \tag{2.4.41}$$

In order to show that

$$|c|_{R_{T(M)}} = \gamma(T(M), \lambda, c) = M \tag{2.4.42}$$

we have to verify the assumptions of Lemma 1.5.10 which then allows to apply Lemma 1.5.11. This in connection with

$$|c|_{R_{T(M)}} = \inf\{|u|_\infty \mid u \in X, S_{T(M)}(u) = c\} = |c|_{T(M)}$$

implies (2.4.42).

Now let $u \in X$ and $t_1, t_2 \in \mathbb{R}$ with $0 < t_1 < t_2 \leq \infty$ be given. Then

$$S_{t_2}(u) = S_{t_2}(u_{t_1}^1) + S_{t_2}(u_{t_1}^2)$$

with

$$u_{t_1}^1 = \begin{cases} u \text{ a.e. on } [0, t_1], \\ 0 \text{ a.e. on } (t_1, \infty], \end{cases} \implies \begin{cases} u_{t_2}^1 \in X \text{ and} \\ |u_{t_1}^1|_\infty \leq |u|_\infty, \end{cases}$$

and

$$u_{t_1}^2 = \begin{cases} 0 \text{ a.e. on } [0, t_1], \\ u \text{ a.e. on } (t_1, \infty]. \end{cases} \implies u_{t_1}^2 \in X.$$

Then

$$S_{t_2}(u_{t_2}^2) = \left(\int_{t_1}^{t_2} e^{-\lambda_j s} u(s) \, ds \right)_{j \in \mathbb{N}} = V_{t_1} S_{t_2 - t_1}(\tilde{u})$$

where

$$\tilde{u}(t) = u(t + t_1) \text{ for all } t \in [0, \infty].$$

Therefore

$$|S_{t_2}(u_{t_1}^2)|_{t_2} = |V_{t_1}S_{t_2-t_1}(\tilde{u})|_{t_2} \le |V_{t_1}|_{B(\ell^1,w)}|S_{t_2-t_1}(\tilde{u})|_{\ell^1}$$

Let $\tilde{t} \in (0,t_1)$ be fixed. Then $V_{t_1} = V_{t_1-t} \circ V_t$ and

$$|V_{t_1}|_{B(\ell^1,w)} \le |V_{t_1-\tilde{t}}|_{B(\ell^1,\ell^1)}|V_{\tilde{t}}|_{B(\ell^1,w)} \le |V_{\tilde{t}}|_{B(\ell^1,w)}$$

As a result we obtain

$$|S_{t_2}(u_{t_1}^2)|_{t_2} \le |V_{\tilde{t}}|_{B(\ell^1,w)}|S_{t_2-t_1}(\tilde{u})|_{\ell^1}$$

where

$$|S_{t_2-t_1}(\tilde{u})|_{\ell^1} = \sum_{k=1}^{\infty}|\int_0^{t_2-t_1} e^{-\lambda_k s}\tilde{u}(s)\,ds$$

$$\le |\tilde{u}|_{\infty}\sum_{k=1}^{\infty}\frac{1}{\lambda_k}(1-e^{-\lambda_k(t_2-t_1)})$$

and

$$\lim_{t_1 \to t_2-0}\sum_{k=1}^{\infty}\frac{1}{\lambda_k}(1-e^{-\lambda_k(t_2-t_1)}) = 0.$$

Thus the assumptions of Lemma 1.5.10 are satisfied and (2.4.42) follows from the above considerations. This in turn implies that

$$|u_M|_{\infty} = M = \gamma(T(M),\lambda,c) \tag{2.4.43}$$

for every $u_M \in X$ with (2.4.41). By virtue of Theorem 2.4.12 this result leads to the following

<u>Theorem 2.4.13</u>: For every $c \in W$, $c \ne \theta_{\ell^1}$ (which implies $c \in S(T,\lambda,c)$ for all $T \in (0,\infty]$) and every $M > \gamma(\infty,\lambda,c)$ there exists $u_M \in S(T(M),\lambda,c)$ with $|u_M|_{\infty} \le M$ and $T(M) > 0$ being defined by (2.4.40) and every u_M satisfies (2.4.43). Moreover, there exists a function $w \in L_1[\varepsilon,T(M)]$ for all $\varepsilon \in (0,T(M))$ which is independent of ε and satisfies (2.4.36) such that for every u_M we have

$$u_M(t) = M \,\text{sgn}\, w(t) \quad \text{for almost all } t \in [0,T(M)]. \tag{2.4.44}$$

This implies that the restriction of u_M to the interval $[0,T(M)]$ is unique.

2.4.4.2. A Direct Approach.

Theorem 2.4.13 can also be proved for $c \in R_\infty$ instead of $c \in W$ by a direct approach as follows:

Let $u_M \in X$ with (2.4.41) be given. Since we assume $c \in R_\infty$ and $M > \gamma(\infty, \lambda, c)$, there exists some $v \in X$ with

$$\int_0^\infty v(t) e^{-\lambda_j t} \, dt = c_j \text{ for all } j \in \mathbb{N}$$

and

$$|v|_\infty < M.$$

This implies that

$$\int_0^{T(M)} (u_M(t) - v(t)) e^{-\lambda_j t} \, dt = \int_{T(M)}^\infty v(t) e^{-\lambda_j t} \, dt = \hat{d}_j, \quad j \in \mathbb{N}$$

where

$$\hat{d} = (\hat{d}_j)_{j \in \mathbb{N}} \in V \subset W.$$

Now we consider the set

$$C = \{ d \in W \mid d = S_{T(M)}(u-v) \text{ for some}$$
$$u \in X \text{ with } |u|_\infty \leq M \}$$

which is convex and closed in W. The convexity is clear and the closedness can be justified as follows. Since the closedness of convex sets holds in all norm-topologies, if it does so in one, it suffices to prove the closedness of C with respect to the ℓ^1-norm-topology. So let a sequence $(d^k)_{k \in \mathbb{N}}$ in C be given such that $\lim_{k \to \infty} |d^k - d|_{\ell^1} = 0$ for some $d \in W$. Then

$$d^k = S_{T(M)}(u^k - v) \text{ for all } k \in \mathbb{N}$$

with $u^k \in X$ and $|u^k|_\infty \leq M$.

Since the set

$$U_M = \{ u \in X \mid |u|_\infty \leq M$$

is weak* sequentially compact there is a subsequence (u^{k_i}) and some $u \in U_M$ such that $u^{k_i} \overset{*}{\to} u$. Since $S_{T(M)}$ maps weak* convergent sequences into norm-convergent ones, as seen above, it follows that for $d = S_{T(M)}(u-v)$ we can conclude that $d \in C$ because of

$$\lim_{i \to \infty} |d^{k_i} - d|_{\ell^1} = 0.$$

Obviously, $\Theta_{\ell^1} \in C$ because of $v \in X$ and $|v|_\infty < M$. In order to see that $\Theta_{\ell} \in \overset{\circ}{C} = $ interior of C we put $\delta = \lambda(T(M))^{-1}(M - |v|_\infty)$ where $\lambda(T(M)) > 0$ such that

$$|d|_{R_{T(M)}} \leq \lambda(T(M)) \, |d|_{R_\infty} \quad \text{for all } d \in R_\infty.$$

Then, for every $d \in W$ with $|d|_{R_\infty} < \delta$, there exists some $u \in X$ with $|u|_\infty < M - |v|_\infty$ and $S_{T(M)}(u) = d$ which implies $S_{T(M)}(u+v-v) = d$ with $|u+v|_\infty < M$, i.e., $d \in C$.

Next we assert, that $\hat{d}$ is on the boundary of C. If this were not the case, i.e., $\hat{d} \in C$, then there would exist some $r \in (0,1)$ such that $\frac{1}{r}\hat{d} \in C$ since $\Theta_{\ell^1} \in \overset{\circ}{C}$. By the definition of C there exists $u_r \in X$ with $|u_r|_\infty \leq M$ and $S_{T(M)}(u_r-v) = \frac{1}{r}\hat{d}$. This implies $S_{T(M)}(ru_r+(1-r)v-v)=\hat{d}$ where $ru_r + (1-r)v \in X$ with $|ru_r + (1-r)v|_\infty < M$. This contradicts the fact that (2.4.43) holds for all $u_M \in X$ with (2.4.41). By a well-known separation theorem for convex sets the existence of some $y^* \in W^*$, $y^* \neq \Theta_{W^*}$ can be concluded such that

$$y^* S_{T(M)}(u_M-v) \geq y^* S_{T(M)}(u-v)$$

for all $u \in X$ with $S_{T(M)}(u-v) \in W$ and $|u|_\infty \leq M$.

Now let $u \in X$ with $|u|_\infty \leq M$ be given. For each $\varepsilon \in (0,T(M))$ we then define

$$u_\varepsilon = (1-\chi_\varepsilon)u_M + \chi_\varepsilon u$$

where

$$\chi_\varepsilon(t) = \begin{cases} 0 & \text{for } t \in (0,\varepsilon) \\ 1 & \text{for } t \in [\varepsilon,T(M)] \end{cases}$$

and obtain $u_\varepsilon \in X$ with $|u|_\infty \leq M$ and

$$S_{T(M)}(u_\varepsilon-v) = S_{T(M)}(u_M-v) - S_{T(M)}(\chi_\varepsilon(u_M-u)).$$

From $S_{T(M)}(u_M-v) = \hat{d} \in W$ and $S_{T(M)}(\chi_\varepsilon(u_M-u)) \in W$ it follows that

$S_{T(M)}(u_\epsilon - v) \in W$ and therefore

$$y^* S_{T(M)}(x_\epsilon(u_M - u)) = y^* S_{T(M)}(u_M - v) - y^* S_{T(M)}(u_\epsilon - v) \geq 0.$$

From Theorem 2.4.11 we deduce the existence of a function $w \in L_1[0, T(M)]$ for every $\epsilon \in (0, T(M))$ which is independent of ϵ, satisfies (2.4.36) and

$$\int_\epsilon^{T(M)} w(t)(u_M(t) - u(t)) \, dt \geq 0$$

for every $u \in X$ with $\|u\|_\infty \leq M$.

By virtue of the proof of Theorem 2.4.12 we are finally led to the

__Theorem 2.4.14:__ Let $c \in R_\infty$, $c \neq \theta_{\ell^1}$, and $M > \lambda(\infty, \lambda, c)$ be given. Then there exists a $u_M \in X$ with (2.4.41) and a function $w \in L_1[\epsilon, T(M)]$ for every $\epsilon \in (=, T(M))$ which is independent of ϵ, satisfies (2.4.36) and

$$u_M(t) = M \, \text{sgn} \, w(t) \quad \text{for almost all} \quad t \in [0, T(M)]$$

for every $u_M \in X$ with (2.4.41). This implies that the restriction of u_M to $[0, T(M)]$ is unique.

2.5. Bibliographical Remarks and References.

The problems of distributed and boundary control for the one-dimensional parabolic partial differential equation (2.1.1) being introduced in Section 2.1 have also been treated by Fattorini and Russell in [2]. They also show that these problems can be reduced to exponential moment problems of the form (2.4.20). However, they take a different approach to the solution of (2.4.20). In order to construct a solution $u \in L_2[0,T]$ they try to find a sequence $(x_j)_{j \in \mathbb{N}}$ in $L_2[0,T]$ which is orthonormal to the sequence $(e^{-\lambda_j t})_{j \in \mathbb{N}}$ and for which the series $\sum_{k=1}^\infty |c_k| \|x_k\|_{L_2[0,T]}$ converges. If such a sequence can be found, then it is immediately clear that $u = \sum_{k=1}^\infty c_k x_k$ is in $L_2[0,T]$ and solves (2.4.20). Starting with the fundamental assumption (2.4.2) they first observe that the sequence $(e^{-\lambda_j t})_{j \in \mathbb{N}}$ is minimal for every $T > 0$. This is not quite correct in view of the Corollary to Theorem 2.4.3 where also condition (2.4.9) is needed. The minimality of the sequence $(e^{-\lambda_j t})_{j\in\mathbb{N}}$ in $L_2[0,T]$ guarantees the existence of an orthonormal sequence $(x_j)_{j\in\mathbb{N}}$ in $L_2[0,T]$ as being shown in Theorem 1.2.5. By the same construction as in the proof of Theorem 1.2.5 they determine a sequence $(x_j)_{j\in\mathbb{N}}$ in $L_2[0,T]$ of the form $x_j(t) = (e^{-\lambda_j t} - r_j(t))/d_j(T)^2$ for $j\in\mathbb{N}$ where $d_j(T)$ is the shortest distance of $e^{-\lambda_j t}$ from the closure $A_j(\lambda)$ of the span of $\{e^{-\lambda_k t} \mid k \neq j\}$ and r_j is the function in $A_j(\lambda)$ which is the closest to $e^{-\lambda_j t}$. On using results by Kaczmarc and Steinhaus [5] on an explicit representation of $d_j(T)$, $j\in\mathbb{N}$, for $T = \infty$ in terms of infinite products only involving the sequence $(\lambda_j)_{j\in\mathbb{N}}$, in connection with Theorem 2.4.2 (where again condition (2.4.9) is needed), they obtain upper bounds for the norms $\|x_j\|_{L_2[0,T]} = \dfrac{1}{d_j(t)}$ for $j\in\mathbb{N}$. These upper bounds allow to prove the following result:

Let $\beta > 1$ and let exist constants $K > 0$ and $\alpha > 0$ such that $\lambda_j = K(j+\alpha)^\beta + 0(j^{\beta-1})$ as $j \to \infty$, then the moment problem (2.4.20) has a solution $u\in L_2[0,T]$ of the form $u = \sum\limits_{k=1}^{\infty} c_k x_k$ where $\sum\limits_{k=1}^{\infty} |c_k| \|x_k\|_{L_2[0,T]} < \infty$, if for some $\eta > 0$

$$\sum_{k=1}^{\infty} |c_k| \exp\{[K^{-1/\beta} (P_\beta - Q_\beta) + \eta] \lambda_k^{1/\beta}\} < \infty$$

where P_β and Q_β are constants with $P_\beta > Q_\beta$, $P_2 - Q_2 = \pi$, $\lim\limits_{\beta\to\infty} P_\beta - Q_\beta = 0$ and $\lim\limits_{\beta\to 1+} P_\beta - Q_\beta = \infty$.

This result contains Theorem 2.4.9 in the special case $\beta = 2$.

In [3] Fattorini and Russell have proved (2.4.25) which leads to Theorem 2.4.8 under the same assumptions as in this theorem, however, with the difference that they take $L_2[0,\infty]$ instead of $L_\infty[0,\infty]$. In addition they make quantitative statements about the function $K(\alpha)$ which is described in terms of properties of the sequence $(\lambda_j)_{j\in\mathbb{N}}$.

In connection with the question of null-reachability of steady states for certain parabolic differential equations in higher space dimensions Gal'chuk considers in [4] moment problems of the form

$$\int_0^T e^{-\lambda_k(T-S)} u(s)\ ds = \frac{\alpha}{\lambda_k}\ ,\ k \in \mathbb{N}, \tag{2.4.45}$$

which we have treated as a special case of the Corollary to Theorem 2.4.7. He shows that, under the condition (2.4.2), for each $\alpha \in \mathbb{R}$ with $|\alpha| < 1$, there exists a time $T > 0$ and a solution $u \in L_\infty[0,T]$ of (2.4.45) with $|u| \le 1$ a.e. on $[0,T]$. This result can be obtained by applying Theorem 2.4.6 as we have shown. This approach follows Schmidt [9]. For the case of one space dimension Krabs and Sachs show in [7] that constant targets are null-reachable by applying Theorem 2.4.4 directly. Theorem 2.4.7 and its Corollary have first been proved by Sachs and Schmidt in [8].

In the survey paper [6] it is also shown how certain problems of distributed and boundary control for parabolic equations in higher space dimensions can be transformed into exponential moment problems of the form (2.4.20). Under the assumptions (2.4.2) and (2.4.9) it is proved that $\|u\|_{\infty,T(M)}$ for each $u \in S(T(M),\lambda,c)$ (2.4.23) with $\|u\|_{\infty,T(M)} \le M$ and $T(M)$ being the minimum time defined by (2.4.40). This implies that $\gamma(T(M),\lambda,c) = M$ where $\gamma(T,\lambda,c)$ is the minimum norm defined by (2.4.31). The proof of controls $\hat{u} \in S(T,\lambda,c)$ with $\|\hat{u}\|_{\infty,T} = \gamma(T,\lambda,c)$ given in [6] is not quite correct. This leads to the rather complicated proof of Theorem 2.4.12. As a result we obtain a bang-bang property and the uniqueness for time-minimal controls. The direct approach to this result given in Section 2.4.4.2 is modelled after the paper [10] by Schmidt.

References.

[1] Dunford, N., and Schwartz, J.T.: Linear Operators. Part II: Spectral Theory. Interscience Pub. Co.: New York 1963.

[2] Fattorini, H.O., and Russell, D.L.: Exact Controllability Theorems for Linear Parabolic Equations in One Space Dimension. Arch. Rat. Mech. Anal. 4 (1971), 272-292.

[3] Fattorini, H.O., and Russell, D.L.: Uniform Bounds on Biorthogonal Functions for Real Exponentials with an Application to the Control Theory of Parabolic Equations. Quart. Appl. Math. 32 (1974), 45-69.

[4] Gal'chuk, L.J.: Optimal Control of Systems Described by Parabolic Equations. SIAM J. Control 7 (1969), 546-558.

[5]   Kaczmarc, S. und Steinhaus, H.: Theorie der Orthogonal-
      reihen. Monografje Matematyczne, Tom VI., Warsaw - Lwow
      1935.

[6]   Krabs, W.: Optimal Control of Processes Governed by Partial
      Differential Equations. Part I: Heating Processes. ZOR 26
      (1982), 21-48.

[7]   Krabs, W., and Sachs, E.: Controllability of Distributed
      Parameter Systems. ZAMM 59 (1979), T103-T105.

[8]   Sachs, E., and Schmidt, E.H.P.G.: On Reachable States in
      Boundary Control for the Heat Equation and an Associated
      Moment Problem. Appl. Math. Optim. 7 (1989), 225-232.

[9]   Schmidt, E.H.P.G.: Boundary Control for the Heat Equation
      with Steady State Targets. SIAM J. Control and Optimization
      18 (1980), 145-154.

[10] Schmidt, E.H.P.G.: The "Bang-Bang"-Principle for the Time-
      Optimal Problem in Boundary Control of the Heat Equation.
      SIAM J. Control and Optimization 18 (1980), 101-107.

[11] Schwartz, L.: Etude de Somme D'exponentielles. Hermann:
      Paris 1959.

Lecture Notes in Control and Information Sciences

Edited by M. Thoma and A. Wyner

Lecture Notes in Control and Information Sciences

Edited by M. Thoma and A. Wyner

Lecture Notes in Control and Information Sciences

Edited by M. Thoma and A. Wyner